"Intelligent Music Production *focuses on the very timely topic of automatic mixing of audio signals. With many clear figures and surprisingly few mathematical formulas, this book explains how artificial intelligence can help producers and musicians make their recordings sound great – automatically.*"

Vesa Välimäki *(Professor of Audio Signal Processing), Aalto University, Acoustics Lab, Espoo, Finland*

T0139332

AUDIO ENGINEERING SOCIETY PRESENTS...

www.aes.org

Editorial Board

Chair: Francis Rumsey, Logophon Ltd.
Hyun Kook Lee, University of Huddersfield
Natanya Ford, University of West England
Kyle Snyder, University of Michigan

Handbook for Sound Engineers
Edited by Glen Ballou

Audio Production and Critical Listening
Jason Corey

Recording Orchestra and Other Classical Music Ensembles
Richard King

Recording Studio Design
Philip Newell

Modern Recording Techniques
David Miles Huber

Immersive Sound
The Art and Science of Binaural and Multi-Channel Audio
Edited by Agnieszka Roginska and Paul Geluso

Hack Audio
An Introduction to Computer Programming and Digital Signal Processing in MATLAB
Eric Tarr

Loudspeakers
For Music Recording and Reproduction
Philip Newell and Keith Holland

Beep to Boom
The Development of Advanced Runtime Sound Systems for Games and Extended Reality
Simon N Goodwin

Intelligent Music Production
Brecht De Man, Ryan Stables, and Joshua D. Reiss

For more information about this series, please visit: www.routledge.com/Audio-Engineering-Society-Presents/book-series/AES

Intelligent Music Production

Intelligent Music Production presents the state of the art in approaches, methodologies and systems from the emerging field of automation in music mixing and mastering. This book collects the relevant works in the domain of innovation in music production, and orders them in a way that outlines the way forward: first, covering our knowledge of the music production processes; then by reviewing the methodologies in classification, data collection and perceptual evaluation; and finally by presenting recent advances in introducing intelligence in audio effects, sound engineering processes and music production interfaces.

Intelligent Music Production is a comprehensive guide, providing an introductory read for beginners, as well as a crucial reference point for experienced researchers, producers, engineers and developers.

Brecht De Man is cofounder of Semantic Audio Labs and previously worked at Queen Mary University of London and Birmingham City University. He has published, presented and patented research on analysis of music production practices, audio effect design and perception in sound engineering.

Ryan Stables is an Associate Professor of Digital Audio Processing at the Digital Media Technology Lab at Birmingham City University, and cofounder of Semantic Audio Labs.

Joshua D. Reiss is a Professor of Audio Engineering at the Centre for Digital Music at Queen Mary University of London.

Intelligent Music Production

Brecht De Man, Ryan Stables and Joshua D. Reiss

NEW YORK AND LONDON

First published 2020
by Routledge
52 Vanderbilt Avenue, New York, NY 10017

and by Routledge
2 Park Square, Milton Park, Abingdon, Oxon OX14 4RN

Routledge is an imprint of the Taylor & Francis Group, an informa business

Library of Congress Cataloguing in Publication Data
A catalog record has been requested for this book

ISBN: 978-1-138-05518-6 (hbk)
ISBN: 978-1-138-05519-3 (pbk)
ISBN: 978-1-315-16610-0 (ebk)

Typeset in Minion
by Newgen Publishing UK

Contents

Figures

Tables

Preface

Organizing this book presented many challenges. Much of the material has been presented in some way before, but always for a different purpose. We have aimed to tell a complete story, in three sections. The first section of the book, Chapters 1 to 3, is primarily concerned with establishing what is already known about music production. Obviously, this is a vast subject, so we focus on the big picture: paradigms, workflows, classifications and so on. Chapter 1 introduces the subject of Intelligent Music Production (IMP), defines the field and gives a summary of its history. Chapter 2 presents audio effects in a consistent framework, discussing the main effects for which intelligent implementations are later described. Chapter 3 covers the theory and approaches to audio mixing, the primary area of music production discussed in this text.

The second section, Chapters 4 to 6, is concerned with how one constructs IMP structures. Chapter 4 first introduces the building blocks behind such systems. It also identifies the key missing element: a body of knowledge concerning the rules and best practices that intelligent systems should follow. The next two chapters address that gap in our knowledge. Chapter 5 describes the means and tools for data collection, and discusses the ways in which music production data can be represented. Chapter 6 details the methods and tools for perceptual evaluation, which both informs development and allows us to assess the performance of IMP techniques.

Chapters 7 to 9 comprise the third section, and are focused on how IMP is performed, in an attempt towards a near-exhaustive overview of the technologies that have been developed. Chapter 7 focuses on IMP systems, primarily those which give intelligence to well-known audio effects, both when operating on their own or in the context of mixing multitrack content. Chapter 8 is about IMP processes, where the focus is not so much on a particular effect, but on how an intelligent system may solve a complex music production task. Chapter 9 deals with IMP interfaces, often designed to give new intuitive, assistive or perceptually relevant tools to users.

The text ends with conclusions and final thoughts in Chapter 10, which also highlights many of the most promising and exciting research trends in the near future. The references are extensive. This is intentional, since we aimed for a comprehensive overview. We hope that this text will give a useful introduction to almost all prior published work in the field.

Acknowledgments

This book describes the work of a large number of researchers who have defined, shaped and advanced the field.

There are far too many to mention here, but we would like to give particular thanks to Dan Dugan, Vincent Verfaille, Bruno Fazenda, Hyunkook Lee, Vesa Välimäki, Bryan Pardo, Richard King and Amandine Pras.

Joshua Reiss is a member of the Centre for Digital Music at Queen Mary University of London. This visionary research group has promoted adventurous research since its inception, and he is grateful for the support and inspiration that they provide. In particular, he supervised or co-supervised Enrique Perez Gonzalez, Alice Clifford, Pedro Pestana, Stuart Mansbridge, Zheng Ma, Dominic Ward and David Ronan, all of whose work is discussed herein. It would be more appropriate though, to say that they needed little supervision and they deserve the credit for the research results that arose.

Ryan Stables is part of the Digital Media Technology lab at Birmingham City University. This book is supported by the research of a number of collaborators from the Digital Media Technology lab including Sean Enderby, Nicholas Jillings, Spyridon Stasis, Jason Hockman, Yonghao Wang, and Xueyang Wang.

Brecht De Man is cofounder of Semantic Audio Labs and previously worked at the Centre for Digital Music and the Digital Media Technology Lab. Writing this book has been made possible by the support of the great people at each of these institutions.

Much of the research described herein was first published in conventions, conferences or journal articles from the Audio Engineering Society (AES). We are greatly indebted to the AES, which has been promoting advances in the science and practice of audio engineering since 1948.

Many discussions have helped shape this book, notably with Kirk McNally and Sean Enderby.

Finally, the authors dedicate this book to their families: Brecht to Yasmine, Nora and Ada; Ryan to Pauline; and Joshua to Sabrina, Eliza and Laura.

Part I
What Do We Already Know?

1
Introduction

"Shannon wants to feed not just data to a Brain, but cultural things! He wants to play music to it!"

Alan Turing (father of modern computing) about
Claude Shannon (father of information theory), during
his 1943 visit to Bell Labs

In this chapter, we give an overview of the history and the state of the art in intelligent audio and music production, with a particular emphasis on the psychoacoustics of mixing multitrack audio, automatic mixing systems and intelligent interfaces.

1.1 Intelligent Music Production – An Emerging Field

Recent years have seen the emergence of intelligent systems aimed at algorithmic approaches to mixing multitrack audio with only minimal intervention by a sound engineer. They use techniques from knowledge engineering, psychoacoustics, perceptual evaluation and machine learning to automate many aspects of the music production process.

An increasing number of companies – from startups to major players in audio software development – have released new products featuring high-level control of audio features, automatic parameter setting, and full 'black-box' music production services. Many of the recent conferences, conventions and workshops by the Audio Engineering Society have featured dedicated sessions on topics like semantic music production or intelligent sound engineering. All of this leaves little doubt about the importance, in both academia and industry, of the wider field of analysis and the automation of music production processes. Existing approaches to problems in these key areas are underdeveloped, and our understanding of underlying systems is limited.

For progress towards intelligent systems in this domain, significant problems must be overcome that have not yet been tackled by the research community. Most state of the art audio signal processing techniques focus on single channel signals. Yet multichannel or multitrack signals are pervasive, and the interaction between channels plays a critical role in audio production quality. This issue has been addressed in the context of audio source separation research, but the challenge in source separation is generally dependent on how the sources were mixed, not on the respective content of each source. Multichannel signal processing techniques are well-established, but they are usually concerned with extracting information about sources from several received signals, and not necessarily about the facilitation or automation of tasks in the audio engineering pipeline, with the intention of developing high-quality audio content.

Thus a new approach is needed. Our goal in most music production tasks relates to the manipulation of the content, not recovering the content. Intelligent Music Production (IMP) has introduced the concept of multitrack signal processing, which is concerned with exploiting the relationships between signals in order to create new content satisfying some objective. Novel, multi-input multi-output audio signal processing methods are required, which can analyze the content of all sources to then improve the quality of capturing, altering and combining multitrack audio.

This field of research has led to the development of innovative new interfaces for production, allowing new paradigms of mixing to arise. However, it has also uncovered many gaps in our knowledge, such as a limited understanding of best practices in music production, and an inability of formal auditory models to predict the more intricate perceptual aspects of a mix. Machine learning has shown great potential for filling in these gaps, or offering alternative approaches. But such methods often have the requirement of being tailored towards problems and practical applications in the domain of audio production.

The following sections present an overview of recent developments in this area.

1.2 Scope

The music lifecycle, from creation to consumption, consists loosely of composition, performance, recording, mixing, mastering, distribution and playback. For the initial music creation stages of the workflow, generative music and algorithmic composition have shown the potential of autonomous music creation systems. Papadopoulos et al. [1] provide an excellent review of twentieth century approaches, and more recently there has been a surge in generative and automatic music composition for everything from elevator music to advertising. It is highly likely that this will become an increasing part of casual music consumption. The latter stages of the workflow, related to music consumption, have already been transformed. On the distribution side, musicians can share their own content at very little cost and effort, and intelligent recommendation systems can find preferred content and create bespoke playlists based on the user's listening history, environment and mood.

1.2.1 Intelligent

By intelligent, we mean that these are expert systems that perceive, reason, learn and act intelligently. This implies that they must analyze the signals upon which they act, dynamically adapt to audio inputs and sound scenes, automatically configure parameter settings, and exploit best practices in sound engineering to modify the signals appropriately. They derive the processing parameters for recordings or live audio based on features extracted from the audio content, and based on objective and perceptual criteria. In parallel, intelligent audio production interfaces have arisen that guide the user, learn their preferences and present intuitive, perceptually relevant controls.

1.2.2 Music

Many of the concepts described herein might also be widely applicable in other audio production tasks. For instance, intelligent mixing technologies could have strong relevance to game audio, where a large number of sound sources need to be played simultaneously and manipulated interactively, and there is no human sound engineer in the games console. Similarly, they are relevant to film sound design, where Foley, dialog and music all need to be mixed, and low budget film and TV productions rarely have the resources to achieve this at a very high standard. However, to keep focus, we assume that the problems and applications are restricted to music.

1.2.3 Production

While there is overlap, we are not specifically referring to music creation – the composition and performance. Nor are we concerned with the distribution and consumption that happens after production. The middle stages of the music workflow – recording, mixing and mastering – are all about the production of the music. They are concerned about how the creative content should be captured, edited and enhanced before distribution.

1.3 Motivation and Justification

The democratization of music technology has allowed musicians to produce music on limited budgets, putting decent results within reach of anyone who has access to a laptop, a microphone, and the abundance of free software on the web [3, 4]. Despite this, a skilled mix engineer is still needed in order to deliver professional-standard material [5].

Raw, recorded tracks almost always require a considerable amount of processing before being ready for distribution, such as balancing, panning, equalization, dynamic range compression, and artificial reverberation to name a few. Furthermore, an amateur musician or inexperienced recording engineer will often cause sonic problems while recording. As noted by Bromham [6], "the typical home studio is entirely unsuitable for mixing records, so there is a greater need than ever to grasp how acoustics will impact our environment and how to work around these inherent shortcomings." Uninformed microphone placement, an unsuitable recording environment, or simply a poor performance or instrument further increases the need for an expert mix engineer [7].

In live situations, especially in small venues, the mixing task is particularly demanding and crucial, due to problems such as acoustic feedback, room resonances and poor equipment. In such cases, having a competent operator at the desk is the exception rather than the rule.

These observations, described in further detail in [114], indicate that there is a clear need for systems that take care of the mixing stage of music production for live and recording situations. By obtaining a high-quality mix quickly and autonomously, home recording becomes more affordable, smaller music venues are freed from the need for expert operators for their front of house and monitor systems, and musicians can increase their productivity and focus on the creative aspects of music production.

Professional audio engineers are often under pressure to produce high-quality content quickly and at low cost [8]. While they may be unlikely to relinquish control entirely to autonomous mix software, assistance with tedious, time-consuming tasks would be highly beneficial. This can be implemented via more powerful, intelligent, responsive, intuitive algorithms and interfaces [9].

Throughout the history of technology, innovation has been met with resistance and skepticism, in particular from professional users who fear seeing their roles disrupted or made obsolete. Music production technology may be especially susceptible to this kind of opposition, as it is characterized by a tendency towards nostalgia [6, 10], and it is concerned with aesthetic value in addition to technical excellence and efficiency. The introduction of artificial intelligence into the field of audio engineering has prompted an outcry from practitioners who reject the concept of previously manual components of their jobs being automated [11].

However, the evolution of music is intrinsically linked to the development of new instruments and tools, and essentially utilitarian inventions such as automatic vocal riding, drum machines, electromechanical keyboards and digital pitch correction have been famously used and abused for creative effect. These advancements have changed the nature of the sound engineering profession from primarily technical to increasingly expressive [12]. Generally, there is economic, technological and artistic merit in exploiting the immense computing power and flexibility that today's digital technology affords, to venture away from the rigid structure of the traditional music production toolset. As more and more industries are starting to consider what implications AI may have (or is already having) for those who work in it, it is opportune to describe the state of the art and offer suggestions about which new tools may become part of the sound engineer's arsenal.

Rapid growth in the quantity of unprocessed audio material has resulted in a similar growth in the engineering tasks and requirements that must be addressed. In audio production for live sound or broadcast one typically has many different sources, each one represented on a separate channel, and they each need to be heard simultaneously. But they could each have been created in different ways, in different environments, with different loudness. Some sources may mask each other, some may be too loud or too quiet, and some may blend in well with the others most of the time, but then have periods where they sound terrible. The final mix should generally have all the sources sound distinct from each other yet contribute to an aesthetically pleasing final product. Achieving this is very labor-intensive, and requires the skills and experience of a professional sound engineer. However, as music content is created by people with a wide variety of backgrounds, there is a need for non-expert audio operators and musicians to be able to achieve a quality mix with minimal effort.

Although production tasks are challenging and technical, much of the initial work follows established rules and best practices. Yet multitrack audio content is still often manipulated 'by hand,' using no computerized signal analysis. This is a time-consuming process, and prone to errors. Only if time and resources permit do sound engineers then refine their choices to produce a mix which best captures an intended style or genre.

1.3.1 Camera Comparison and the 'Instamix'

Professional music production systems offer a wide range of audio effects and processors. But they all require manual manipulation. In effect, as technology has grown, functionality has advanced, but it has not become simpler for the user.

In contrast, the modern digital camera comes with a wide range of intelligent features to assist the user. These include face, scene and motion detection, autofocus and red eye removal. An audio recording or mixing device has none of this. It is essentially deaf. It doesn't listen to the incoming audio, and has no knowledge of the sound scene or of its intended use. It generally lacks the ability to make processing decisions, or to adapt to a different room or a different set of inputs. Instead, the user is forced to either accept poor sound quality, or to do a significant amount of manual processing.

Extending the photography analogy, one could imagine desiring automated parameter adjustments while still controlling the overall style, mimicking a certain technique, or evoking a certain era, which is possible when combining a phone camera with an app like Instagram. Similarly, the amateur music producer or hurried professional could apply a 'filter' corresponding with a certain decade, musical genre, or famous sound engineer, though only if these can be quantified as a function of parameter settings or measurable sound properties. For instance, an inexperienced user could produce a mix that evokes a 'classic rock' sound, a 'Tom Elmhirst' sound, or a '1960s' sound, thus providing a starting point for the novice to achieve a creative goal.

1.3.2 Sound on Sound

Another way to describe the motivation for this work can be found in a 2008 editorial in *Sound on Sound* [13]. The magazine's editor wrote, "There's no reason why a band recording using reasonably conventional instrumentation shouldn't be EQ'd and balanced automatically by advanced DAW software." And that's the point: intelligent software should be able to automate many of the decisions made by a sound engineer.

He also discussed "musicians who'd rather get on with making music than get too deep into engineering," which is an important motivation for us too. One of our goals is to address the needs of the musician who doesn't have the time, expertise or inclination to perform all the audio engineering required. Why should a classical pianist, who has spent the past 20 years mastering their instrument, be required to take a course in engineering, simply to produce a commercial recording?

Finally, there was the comment "[audio interfaces can] come with a 'gain learn' mode... DAWs could optimize their own mixer and plug in gain structure while preserving the same mix balance." That is the specific path that has been taken by many (but not all) IMP techniques: to preserve the choices of the engineer, while automating those other, more standard tasks.

1.3.3 Aims and Objectives

Motivated by these challenges, the research described in this book has aimed to make music production systems more intuitive, more intelligent, higher quality, and capable of autonomous operation. The objectives of much of this research may be described as follows:

- Take care of the technical aspects and physical constraints of music production. For example, limit gain to avoid distortion, or reduce spectral masking of sources to improve intelligibility.
- Simplify complex music production tasks. For example, set some dynamic processing parameters based on features extracted from the source audio.
- Provide intuitive and perceptually relevant interfaces, more directly tailored to the needs of the user.
- Allow for fully autonomous operation, thus enabling amateur musicians and performers to produce content when there are not sufficient resources for a professional sound engineer.

It has also sought to investigate the extent to which these objectives can be achieved. Notably, it explores whether a software system can produce an audio mix of comparable quality to a human generated mix.

1.4 History of the Field

1.4.1 Early Automation in Audio Effects

Many of the classic audio effects were designed as a means to automate, or at least simplify, some aspects of music production.

Early equalizers (EQs) were fixed and integrated into audio receivers and phonograph playback systems, but users and their applications demanded control. Tone controls emerged in the middle of the twentieth century, giving the ability to apply boost or cut to bass and treble [14, 15]. It wasn't until the work of Massenburg [16] that the parametric EQ emerged, allowing control that was more aligned with the needs of audio engineers for smooth yet precise adjustment of the spectrum, attuned to the logarithmic scale (in both frequency and amplitude) of human hearing.

The parametric EQ, though versatile, is not necessarily intuitive. It requires training for use. Reed [9] provided a new interface for EQ that could perhaps be considered the first intelligent EQ assistant. Inductive learning based on user feedback was used to acquire expert skills, enabling semantic terms to be used to control timbral qualities of sound in a context-dependent fashion. Similarly, Mecklenburgh and Loviscach [17] trained a self-organizing map to represent common equalizer settings in a two-dimensional space organized by similarity. However, the space was hand-labeled with descriptions of sounds that the authors considered to be intuitive. A detailed overview of these EQ methods can be found in a recent review paper by Välimäki and Reiss [18].

Automation is more emphasized in the evolution of dynamic range compression. Early use of dynamic range compression was often to fit the signal's amplitude range to that of a recording medium. It also served the purpose of automating the continual riding of faders. But the parameters and use of a compressor are far from intuitive. Tyler [19] provided a simplified version of a compressor, but in the process sacrificed its versatility and much of its adaptive nature. In earlier studies [20, 21], peak and RMS measurements were used to automate the selection of time constants. Automatic make-up gain can be found in some compressor designs, but only as a fixed compensation that does not depend on signal characteristics such as loudness, even though the main purpose of make-up gain is to achieve the same loudness between the uncompressed and compressed signals. More perceptually relevant approaches to dynamic range compression, incorporating some measure of loudness or loudness range, emerged early in the new millennium [22, 23].

However, the biggest leap in intelligent production tools occurred in a multitrack context with the simplest of effects: the fader or level control.

1.4.2 Automatic Microphone Mixing

The idea of automating the audio production process, although relatively unexplored, is not new. It has its origins in automatic microphone mixing. The terminology is used loosely, and automatic microphone mixing is often shortened to automixing. But it refers to a restricted form of automation: automatic microphone level control for speech. In standard speech mixing, it is common practice to open only those microphones that are in use. This reduces noise and maximizes gain before feedback.

Automatic Microphone Mixer Objectives

Dan Dugan [24] stated the basics of microphone mixing and showed that every doubling of the number of microphones reduces the available gain before feedback by 3 dB. Most current designs will restrict the maximum number of microphones to be open regardless of the overall number of microphones in the system. In practice, they generally do not go from a complete 'on' to 'off' state, but instead produce only a 15 dB level change, or some user controllable value [25]. The input channel attenuation setting accomplishes the purpose of optimizing the system gain before feedback, especially when the number of microphones is large.

According to the *Handbook for Sound Engineers* [25], the design objectives of an automatic microphone mixer are:

- Keeps the sound system gain below the threshold of feedback.
- Does not require an operator or sound technician at the controls.
- Does not introduce spurious, undesirable noise or distortion.
- Easy installation, like a conventional mixer.
- Is able to adjust the system status outputs for peripheral equipment control and interface with external control systems.
- Responds only to speech signals and remains mostly unaffected by extraneous background noise signals.
- Activates input channels fast enough that no audible loss of speech occurs.
- Allows more than one talker on the system when required, while still maintaining control over the overall sound gain.

The last three objectives are speech-specific, and require modification for use with musical signals. For music mixing, opening and closing inputs may result in unnatural artifacts.

Existing microphone mixers tend to be part of huge conference systems and often interact with other devices using standard communication protocols. Some of these conference systems have multiple rooms and outputs, which adds complexity to the mixer output stages.

Automatic Microphone Mixer Designs

Early automatic microphone mixer systems were only concerned with automatic level handling and required a significant amount of human interaction during setup to ensure a stable operation [25]. One method for controlling the level is use of an automatic gain control, or AGC. The automatic gain control mixers operate by setting up the quietest active microphone as the reference gain. The microphone has maximum gain before feedback while louder talkers will activate the AGC to reduce the overall level. AGC tends to have similar control parameters to a dynamic range compressor, with attack and release time specified to minimize artifacts. This increases the user complexity of an automatic mixer. In many cases the settings are fixed by the manufacturer, which simplifies their operation but limits their application.

Fixed threshold mixers use a gate to open and close the microphone channel. When the microphone input signal exceeds a given threshold, the gate switches from closed to open. The static threshold has the disadvantage of impeding low-level signals. It may cause artifacts or miss the first sections of a word.

Gain sharing designs [24, 26] are based on the premise that the sum of the signal inputs for all active microphones must be below the minimum value that causes feedback. This value is usually set manually by the user.

Directional sensing [27] uses microphones that both capture the source and estimate the ambient noise level. An input is activated only if the source level is significantly above the noise, and only one microphone is opened per active source.

Multivariable dependent mixers take into account both amplitudes and timing of active signals. Initially, all inputs are attenuated. It preserves the relative gains of the speakers since all output gains are the same. If many speakers attempt to talk at the same time the probability of their microphones being open decreases.

Noise-adaptive threshold mixers use dynamic thresholds for each channel. This enables them to distinguish between signals whose frequency content and amplitude is constant, such as air conditioning noise, and signals rapidly changing in frequency and amplitude, such as speech. Other considerations can be added to this design to ensure that a loud talker does not activate multiple inputs.

None of these early systems were designed to work with music. The variable threshold mixer attempted this [24]. It had the ability to adapt the threshold based on a signal from an additional microphone that captures the room noise contributions, a form of adaptive gating. But it was not fully developed, was only capable of controlling levels and was not based on any perceptual attributes.

Ballou [25] summarized the state of the field as follows:

> The operational concepts used in digital automatic microphone mixers have not varied far from the previously described concepts underlying the analog automatic microphone mixers. This is likely to change, but as future digital automatic mixing concepts will be hidden deep within computer code the manufacturers may be unwilling to reveal the details

of operational breakthroughs; they will likely be kept as close guarded company secrets. New concepts in automatic mixing might only become public if patents are granted or technical papers are presented.

This statement identified a lack of development in this field. It also acknowledged that due to an unwillingness to publish or disclose, many of these advancements are likely to pass unnoticed in the scientific community.

1.4.3 Steps Towards Intelligent Systems

In the early 2000s, research towards IMP systems was in its infancy, partly because digital music production was still an emerging field. Around that time, Pachet et al. [28] presented the idea of maintaining the intentions of the composer and sound engineer while providing the end user with a degree of control. This system provided the user with controllable parameters, constrained to keep aesthetic intention. It was designed to work with pre-recorded material and was unable to deal with live musical sources. It also required human programming.

Katayose et al. [29] proposed a 'mix-down assistant,' which might apply a target profile to source content in order to mimic the approach of a certain engineer. Similarly, Dannenberg et al. [30, 31] described an intelligent audio editor, an auto-adaptive environment where, among other things, tracks would be automatically tuned. It described a system that was capable of maintaining a target, user-defined loudness relationship. It was intended as a tool to automatically time-align and pitch-correct performances, which are auto-adaptive implementations that transcend into the automatic editing realm. But it required a machine-readable score, and hence was highly limited in its application.

Due to the huge growth in audio codecs, research during this period also focused on mixtures of large numbers of audio signals where data compression is achieved by removing sounds that are masked by other sources [32–35]. Although relevant for real-time audio applications where data size is an issue, it is only concerned with mixes produced by summing the sources. And data size is not a key challenge addressed in the music production process.

1.4.4 The Automatic Mixing Revolution

Between 2007 and 2010, Enrique Perez Gonzalez, an experienced sound engineer and music technology researcher, gave new meaning to the term by publishing methods to automatically adjust not just level [39], but also stereo panning of multitrack audio [36], equalization [40], delay and polarity correction [72]. He automated complex mixing processes such as source enhancement within the mix (a multitrack form of mirror equalization) [38] and acoustic feedback prevention [37], and performed the first formal evaluation of an automatic mixing system [43]. To our knowledge, this was the inception of the field as it is known today.

This ushered in a burst of sustained activity in the Automatic Mixing field [73], an important subset of IMP. Figure 1.1 shows a comprehensive but not exclusive overview of published systems or methods to automate mixing subtasks during the ten years beginning with Perez's first automatic mixing paper. Some trends are immediately apparent from this timeline. For instance, machine learning methods seem to be gaining popularity [65–70].

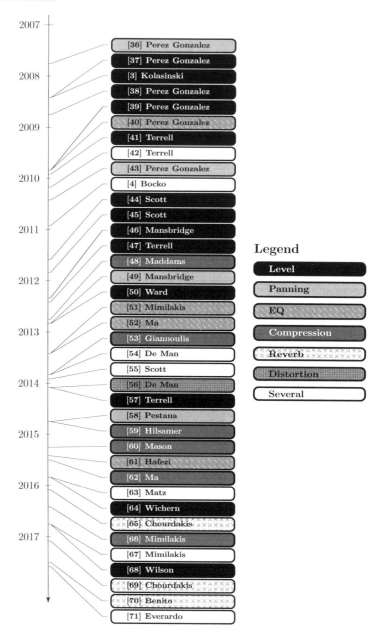

Figure 1.1 Timeline of prior automatic mixing work 2007–2017

Whereas a majority of early automatic mixing systems were concerned with setting levels, recent years have also seen automation of increasingly 'complex' processors such as dynamic range compressors [48, 53, 59, 60, 62, 66] and reverb effects [65, 69, 70]. Research on such systems has additionally inspired several works on furthering understanding of the complex mix process and its perception [12, 74–76].

1.5 Applications and Impact

Currently, most audio processing tools demand manual intervention. For instance, although audio workstations are capable of saving a set of static scenes for later use [77], they lack the ability to take intelligent decisions, such as adapting to different acoustic environments or a different set of inputs.

But with the advent of the technologies described in this book, this is all changing. The purpose of many IMP tools is to reduce the work burden on the engineer, producer or musician, and to explore the extent to which various tasks can be automated. Knowledge about the process of mix engineering has many immediate applications, of which some are explored here.

They range from completely autonomous mixing systems, to more assistive, workflow-enhancing tools. The boundaries between these categories are vague, and most systems can be adapted for less or more user control.

1.5.1 Intelligent Assistants

In 2000, Moorer [78] introduced the concept of an Intelligent Assistant, incorporating psychoacoustic models of loudness and audibility, intended to "take over the mundane aspects of music production, leaving the creative side to the professionals, where it belongs". Adding control over high-level parameters, such as targeted genre, shifts the potential of IMP systems from corrective tools that help obtain a single, allegedly ideal mix, to creative tools offering countless possibilities and the user-friendly parameters to achieve them. This mitigates the risk – and frequent criticism of skeptics – that mixes become generic, as a result of a so-called 'color by numbers' approach lacking creative vision and the intuition of an expert [6].

Sound engineers of varying levels will typically want some degree of control, adjusting a number of parameters of mostly automatic systems. Such a system can quickly provide a starting point for a mix, or reach an acceptable balance during a sound check, like a digital assistant engineer. Even within a single processor, extracting relevant features from the audio and adjusting the chosen preset accordingly would represent a dramatic leap over the static presets commonly found in music production software [3].

Assistants can range from very limited automation – the already common automation of ballistics time constants and make-up gain in dynamic range compressors – to only exposing a handful of controls of a comprehensive mixing system. An example of the latter can be found in online mastering services, where the user often controls as little as a single sonic attribute.

Large audio productions often have hundreds of tracks. Being able to group some of those tracks into an automatic mode, or even automate the grouping, will reduce the effort required by the audio engineer. There is also the possibility of applying this technology to remote mixing applications where latency is too large to be able to interact with all aspects of the mix.

Riders

One attempt to reduce the required effort of the sound engineer while mixing multitrack content is the development of automatic riders. A rider is a type of gain control which continually and smoothly adjusts the gain on a channel to match a given criterion. Recently riders have been devised that, given the desired loudness level of a target channel in relation to the rest of the mix, will compensate for all deviations by raising or lowering the levels of the target [12,79]. The existing riders only work with at most two channels of audio, and thus riders have limited flexibility. All channels other than the target are combined into a single channel, only the target is modified, and only a single loudness curve can be used.

Automating fader riding is a quintessential example of assistive mix technology, as it used to be a tedious and manual job where sometimes several engineers moved a number of faders to change levels dynamically over the course of a song. This task was then automated to some extent by the dynamic range compressor as well as by programmable motorized faders or digital gain automation.

1.5.2 Black Box

In engineering terms, a 'black box' is a system which can only be judged based on its output signals, in relation to the supplied input. In other words, the user does not know what goes on inside, and cannot control it except by modifying the incoming signals. One or more mix tasks could be automated by such a device so that no sound engineer is required to adjust parameters on a live or studio mix [13,47,80].

Many of the academic approaches are presented this way, given the appeal of presenting a fully automatic system, although they could be generalized or only partially implemented to give the user more control. The absence of a need – or option – for user interaction is a desired characteristic of complete mixing and mastering solutions, for instance for a small concert venue without sound engineers, a band rehearsal, or a conference PA system.

Live Music Production Tools

Methods have also been proposed for intelligent *live* music production systems. Some of these tools take advantage of mixing board recall functionality and add autonomous decisions to the mixing process. Much of the research described here was for live sound applications. But to our knowledge, no commercially available mixing device is yet capable of equalizing and organizing the gain structure while taking care of acoustic or technical constraints.

1.5.3 Interfaces

Another class of IMP tools, complementary to Automatic Mixing in the strict sense, comprises more or less traditional processors controlled in novel ways. For instance, a regular equalizer can be controlled with more semantic and perceptually motivated parameters, such as 'warm', 'crisp' and 'full' [81, 375], increasing accessibility towards novices and enhancing the creative flow of professionals.

Deviating from the usual division of labor among signal processing units, the control of a single high-level percept can be achieved by a combination of equalization, dynamic range compression, harmonic distortion or spatial processing. Such intuitive interfaces are likely to speed up music production tasks compared to traditional tools, but also facilitate new ways of working and spur creativity. In the pro-audio realm, adaptive signal processing and novel interfaces allow users to try out complex settings quickly [17].

Already, music software manufacturers are releasing products where the user controls complex processing by adjusting as little as one parameter. New research is needed to validate these relationships, uncover others, and confirm to what extent they hold across different regions and genres.

1.5.4 Metering and Diagnostics

Intelligent metering constitutes another possible class of systems built on this new information, taking the omnipresent loudness meters, spectral analyzers, and goniometers a step further, towards more semantic, mix-level alerts [410]. For instance, the operator can be warned when the overall reverb level is high [82] or the spectral contour too 'boxy.' By defining these high-level attributes as a function of measurable quantities, mix diagnostics become more useful and accessible to both experts and laymen. Such applications also present opportunities for education, where aspiring mix engineers can be informed of which parameter settings are considered extreme. Once these perceptually informed issues have been identified, a feedback loop could adjust parameters until the problem is mitigated, for instance turning the reverberator level up or down until high-level attribute 'reverb amount' enters a predefined range.

1.5.5 Interactive Audio

Growth in demand for video games and their ever-increasing audio processing requirements makes Intelligent Audio Production for games a promising research area. In game or interactive audio, hundreds of ever changing audio stems need to be prioritized and mixed in real time to enhance the user experience. Virtual reality and game audio often have dynamic environments where users may change their location with respect to the sources, and sources are then rendered to give the appropriate spatial characteristics [28]. This suggests further constraints; some sources have a predetermined spatial position and others can be adjusted. Any automation or streamlining of this process could be beneficial, especially since the automation is not 'competing' with a professional, creative sound engineer. That is, the challenge is to either perform Intelligent Audio Production, or not perform anything except a simple sum of sources with some basic rules applied.

2
Understanding Audio Effects

In this chapter, we give an overview of audio effects. We describe their mathematical formalism, several schemes for their classification, the building blocks for their implementation and how they may be imbued with 'intelligence.' The scope of the chapter is quite broad, and discusses audio effects in all of their forms. However, details of specific intelligent implementations of these effects are reserved for Chapter 7.

2.1 Fundamental Properties

Some key mathematical concepts are incredibly useful in describing audio effects. By understanding these properties and whether they apply to a given effect, one can gain immediate intuition about its behavior.

Linear systems obey the linearity principle, $f(ax_1 + bx_2) = af(x_1) + bf(x_2)$. So there is no sonic difference between applying a linear effect to a sum of sources, or applying it to the sources individually before they are summed. Equalizers are typically linear. But dynamic range compression is not linear due to the dependence on signal level. For instance, two inputs may both never reach a level high enough to apply compression, even when their sum is always compressed.

A *time-invariant* system is, quite simply, any system that does not directly depend on time. That is, if the input signal $x(t)$ gives an output signal $y(t)$, then if we delay the input signal by some amount, $x(t - \tau)$, it gives the same output signal, just delayed by that same amount, $y(t - \tau)$. Note that here $x(t)$ represents a signal (technically for $-\infty$ to $+\infty$), not a value at some instant t. The discrete time equivalent of time invariance is shift invariance $(y[n] = f(x[n]) \rightarrow y[n - m] = f(x[n - m]))$. Many common effects are driven by low-frequency oscillators (LFOs). Such effects are not time-invariant, since output depends on the current (explicitly time-dependent) phase of the LFO, but their effect is identical every integer number of LFO periods.

Most introductory signal processing deals with linear, time-invariant (LTI) systems, which includes audio effects like equalization, panning, gain and delay. This is convenient since LTI systems have lots of important properties. For instance, they can be completely described by an impulse response. But standard signal analysis approaches often do not apply to the many audio effects that are nonlinear or time-varying.

Real time does not have a universally accepted definition, and the term has related but different usages for real-time computing, real-time simulation, real-time rendering and so on. Perhaps the most useful definition for our purposes is that a real-time system "controls an environment by receiving data, processing them, and returning the results sufficiently quickly to affect the environment at that time" [83]. In practice, this means that the computing time needed to perform all operations within a time interval must be less than its playback time.

Real-time systems typically operate in discrete time with a fixed time step. But the time step for processing need not be the sample period. For instance, audio could be sampled at 44.1 kHz, but processed in frames of 1024 samples, so the time step for processing is 1024/(44100 Hz), or approximately 23 ms. Also, systems are sometimes described as 'faster than real time.' Such terminology may be deceiving since it often just refers to computational time with certain hardware. For instance, a system may take ten seconds to process a minute of audio, but it could require access to the entire minute of audio before it can begin analysis.

A system can process inputs *online*, i.e., simultaneously returning outputs in a timely manner, or *offline*, i.e., after all inputs have been recorded. In the former case, it has to work in real time or faster, but in the latter case there is no such requirement. As such, a real-time system may well work offline, but a system that is not real time can never work online.

A *causal* system is one where the output depends on past and current inputs (and possibly outputs) but not on future inputs. If there is some dependence on future inputs, it is a non-causal system. An offline system typically has access to all inputs and can therefore let processing of the first piece of data depend on the properties of the last bit.

Latency refers to the delay between when an input signal enters a system and when the processed version of that input emerges at the system output. Latency is critical in many audio applications since multiple performers aim to synchronize what they play, as well as synchronize the sound with what the audience see. We sometimes refer to 'live processing', which is achieved by real-time systems with sufficiently low latency, thus enabling the use of audio effects in performance.

Real-time systems are inherently causal, but may require latency for causality. Some latency allows a real-time system to seem to 'look into the future' a bit before deciding how to process content at a given time. For instance, a pitch shifting technique may require a frame of input before deciding how the first sample within that frame will be processed. In that case, it is causal with regard to frames, but not necessarily with regard to samples.

To minimize latency, *sample-based* processing rather than *frame-based* processing is often preferred for live applications. That is, effects are implemented so that the processing on any given sample is not dependent on future samples: sample in, processed sample out. Such requirements often dictate the type of algorithm that can be applied.

2.2 Audio Effect Classification

There is a bewildering variety of audio effects, which makes any systematic classification of them particularly challenging. But classification, including development of taxonomies and ontologies, is important since it facilitates the choice of the right audio effect for the right application, and also underpins the combination and ordering of multiple audio effects to achieve complex tasks.

Verfaille [84] notes three main ways in which audio effects may be classified:

1. the perceptual attribute that is modified;
2. the implementation technique; or
3. the type of control that is employed;

and further subcategories, which make use of some of the properties described in Section 2.1.

These approaches to classification have been used by others. For instance, classifying audio effects based on the perceptual attributes that are modified (delay, pitch, positions or quality) was the main approach in [85, 86], whereas implementation (delays, filters, modulators, time segment processing, time-frequency processing...) was the distinguishing factor in two textbooks on audio effects [87,88].

The perceptual attributes modified are important for users, whereas developers and programmers are most concerned with the implementation technique. The type of control that the effect employs is relevant both to developers in terms of how they are constructed, and users in terms of how they might interact with the effect. This approach is less widely adopted, but as we shall see, is very useful in establishing a framework for Intelligent Music Production tools.

Below, we delve deeper into these classification schemes.

2.2.1 Classification by Perception

Crucial to any classification concerned with perceptual attributes is the way in which those attributes are distinguished. The approach in [85,86,89] and others all used variations on the following categories.

Pitch and harmony: Pitch can be defined as the frequency of the sinusoid that can be consistently matched by listeners to the sound being analyzed. In practice, manipulation of pitch often involves changes to the fundamental frequency and its harmonics. The combination of pitches into notes, chords and melodies constitute the harmonic aspects, and audio effects aimed at modifying harmony often analyze the high-level musical structure in the signal.

Dynamics: These audio effects aim mainly to adjust the amplitude characteristics, and are most often implemented in the time domain. They often aim to modify perceived loudness or loudness range, but may also adjust nuances, accents, sustain, tremolo and phrasing (legato and pizzicato) in a performance.

Temporal aspects: Time is perceived both through the duration of sound and gaps, and the rhythm, which is based on repetition and inference of patterns. Time-scaling is a well-known example of an effect intended only to modify temporal aspects, though many others seek just to manipulate the rhythm (accelerando, deccelerando, periodic reordering and repetition of fragments).

Spatial aspects: These include attributes such as width, distance, elevation, azimuth, diffuseness, directivity, and room effects such as reverberation and echo.

Timbre: Timbre is a poorly defined perceptual attribute. It is sometimes awkwardly characterized as any characteristic that distinguishes two sounds of the same pitch and intensity when played under the same conditions. That is, it is a catch-all for any perceived qualities that cannot be assigned to the other categories. However, many timbral characteristics are closely associated with the envelope of the spectrum, i.e., the coarse structure as opposed to the fine structure which gives the pitch. As such, they relate to manipulation of statistical audio features such as spectral contrast, spectral flux and spectral skewness.

The emphasis on perception suggests but does not dictate the implementation. For instance, reverberation may be implemented as a single channel effect, even though it changes spatial characteristics.

One way to represent this classification is with a mind map (see Figure 2.1) where the effects are first distinguished by the main perceptual attribute modified. Further refinements may be given by describing secondary attributes that the effects modify.

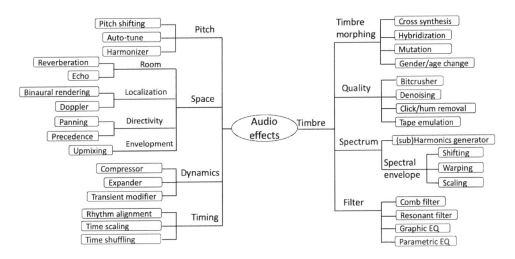

Figure 2.1 A simple map classifying many audio effects by the main perceptual attributes they modify

2.2.2 Classification by Technique

According to [84, 87], digital audio effects can be organized on the basis of implementation techniques along the following lines:

Filters: Typically linear, time-invariant operations, intended to shape the frequency spectrum of a signal.

Delays: Wide range of effects that are typically implemented with delay lines, such as filters, echo, slapback, chorus and flangers.

Modulators and demodulators: Parameters of a signal are modified by another time-varying signal such as a sinusoid (modulation) or where that process is reversed (demodulation).

Nonlinear processing: Both dynamics processing (compressors, limiters, noise gates and expanders) and effects intended to introduce or model nonlinearities (distortion, waveshaping, analog emulation).

Spatial effects: The wide field of spatialization technology, including binaural processing and multichannel reproduction.

Time-segment processing: Typically time domain block-based processing methods, such as Synchronous Overlap and Add (SOLA).

Time-frequency processing: A two-dimensional time-frequency representation of the input audio signal is manipulated.

Source-filter processing: The audio signal is modeled as a source and a filter. These separate aspects can be manipulated to produce cross-synthesis, pitch shifting and other effects. Also known as the excitation-resonance model.

Spectral processing: The audio signal is represented by a spectral model, such as in sinusoidal modeling synthesis (SMS).

Time and frequency warping: Deforming the time and frequency axes of an audio signal.

Classification based on the underlying techniques enables the developer to see the technical and implementational similarities of various effects, thus supporting reuse of code and development of multi-effect systems. However, some audio effects can be classified in more than one class. For instance, time-scaling can be performed with time-segment or time-frequency processing. This classification may not be the most relevant or understandable for audio effect users, nor does this classification explicitly handle perceptual features.

As in Figure 2.2, users may choose between several possible implementations of an effect depending on the artifacts of each effect. For instance, with time-scaling, resampling does not preserve pitch nor formants; OLA (Overlap and Add) with a circular buffer adds the window modulation and sounds rougher and filtered; a phase vocoder sounds a bit reverberant; the 'sinusoidal + noise' additive model sounds good except for attacks; the 'sinusoidal + transients + noise' additive model preserves attacks, but not the spatial image of multichannel sounds; etc. Therefore, in order to choose a technique, the user must be aware of the audible artifacts of each technique. The need to link implementation techniques to perceptual features thus becomes clear.

2.2.3 Classification by Control

Although these classifications are useful in many contexts, they are not ideal for understanding the signal processing control architectures of some more complex effects,

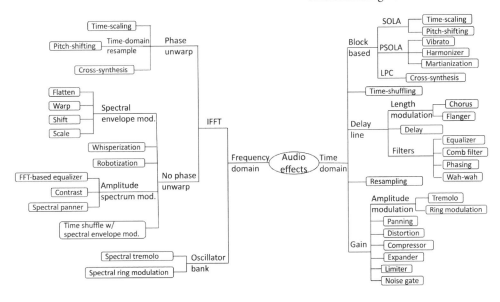

Figure 2.2 A simple map classifying many audio effects by the main technical approach they use.

especially the intelligent effects which we come to later. Thus we use the approach proposed in [89], which classifies digital audio effects in terms of the ways in which control parameters may be determined by input audio signals. At first, this may seem counterintuitive – why would the control parameters, as opposed to the perceptual properties or signal processing techniques, be an important aspect? However, certain forms of control are an essential enabling technology for intelligent audio effects.

Having control parameters depend on an audio signal is what defines an adaptive audio effect. Verfaille et al. [89] further break down adaptive audio effects into several subcategories. In what follows, non-adaptive, auto-adaptive, external-adaptive and cross-adaptive processing devices will be discussed.

Non-adaptive

For non-adaptive audio effects, all parameters that influence the processing are either fixed or directly controlled by the user. In general, features are not and do not need to be extracted from input signals. For instance, a graphic equalizer is non-adaptive. The locations of the frequency bands are fixed and the gain for each band is controlled by the user. No aspect of the processing is dependent on any attributes of the input signal.

Figure 2.3 shows the standard implementation of a non-adaptive audio effect, where n is the discrete time index in samples, $x[n]$ is the input source and $y[n]$ is the output resulting from the signal processing.

Adaptive digital audio effects, in contrast, use features extracted from the audio to control the signal processing. At first glance, the distinction between adaptive and non-adaptive may not be clear-cut. Consider processing the frequency content with an equalizer versus processing the dynamic content with a dynamic range compressor. Both approaches offer a set of user controls which may be fixed or changed during operation. These controls establish the frequency response of the equalizer and the input/output curve of the compressor.

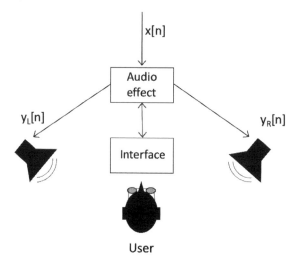

Figure 2.3 Diagram of a non-adaptive audio effect where the user interacts with an interface to control its parameters.

Yet the compressor is adaptive and the equalizer is not. The key to this distinction is an understanding of how they are designed and what that implies. The equalizer applies the same processing to the input signal regardless of the signal content. Even though the output signal depends on the input, it does not need to perform any analysis on the input in order to determine how it will be processed. In contrast, the gain applied by a compressor depends on the input signal level. Thus, the input signal needs to be analyzed and the level determined.

There is a subjective element to this distinction. In some cases, the analysis of the input signal could be incorporated into the processing, blurring the line between adaptive and non-adaptive. Thus, implementation details are an important factor in this classification.

It is also important to note that this distinction between non-adaptive and adaptive effects is not the same as between linear and nonlinear. Consider treating an audio effect as a black box, which is fed sine waves at different frequencies. For a linear, time-invariant effect, the same frequencies occur at the output, only with the amplitude and phase changed. One could imagine an effect that, regardless of input, always adds a 1 kHz tone at the output. Such a system would be non-adaptive, but it would also be nonlinear.

Auto-adaptive

In an auto-adaptive audio effect, the control parameters are based on a feature extracted from the input source itself. Here 'auto' means 'self' and is not an abbreviation of 'automatic,' in contrast to auto-tuning or auto-mixing. In feedforward designs, the analysis and feature extraction is performed only on the input signal. However, auto-adaptive audio effects can also use feedback from the output signal. An example effect that exists in both forms is the dynamic range compressor, where the gain applied to the signal is dependent on a measurement of either the input or output signal level, for feedforward and feedback compressors respectively [90].

In their seminal work on adaptive digital audio effects, Verfaille and Zölzer [89] provided brief descriptions of several auto-adaptive effects. Auto-adaptive audio effects include, for example, time-warping, noise gates and compressors.

Morrell and Reiss [91] presented a creative implementation of an automatic panner where a track's RMS value controls its drift from a static pan position. In a second implementation, the authors substituted RMS features with the spectral centroid, so that the effect became a frequency-dependent panner.

'Corrective track equalization' sees equalization as a way of correcting imperfections in an individual source or recording technique based on analysis of the audio signal, thus being an auto-adaptive process [92, 93].

Dannenberg [30] described a system where tracks would be tuned by automatically adjusting pitch, timing and dynamic level. Such an auto-adaptive environment opens up the possibility of automatic editing.

External-adaptive

In an external-adaptive audio effect, the system derives its control processing variable from analysis of a different source to the one on which it has been applied. This is the case for ducking effects. Whenever the secondary audio input reaches a certain volume level, the audio source being processed 'ducks' out of the way by reducing in volume. The technique can clearly be heard on the radio when a DJ begins to speak and the music automatically reduces in volume.

An external-adaptive effect is feedforward if it takes its control variable only from the input signals, and it is a feedback external-adaptive effect if it relies on analysis of the output signal. As defined, the effect is *not* adaptive to the input source. However, the feedback implementation may indirectly adapt to the input source, given that the output could depend on both the input source and the external source.

Cross-synthesis, for example, is a feedforward external-adaptive effect using the spectral envelope of one sound to modify the spectral envelope of another sound.

The Talk Box effect, which has been popular since the 1970s, can be classified as an external-adaptive effect where the frequency content of the instrument's sound is filtered and modulated by the shape of the performer's mouth before amplification. The traditional Talk Box is an interesting example since it does not use analog or digital circuitry for the analysis and modulation, but rather the physical characteristics of how the mouth processes the human voice and the audio input signal. The Talk Box effect is often confused with the vocoder technique, which is also an example of an external-adaptive effect where the spectral envelope of one audio signal modifies the spectral envelope of another audio signal electronically.

Cross-adaptive

Cross-adaptive audio effects are characterized by having control parameters which are determined by several input audio signals. That is, these effects analyze the signal content of several input tracks (and often the relationships between them) to inform the processing of at least one output track. By removing or adding sections it can conform to any of the previous adaptive topologies. It generalizes the single track adaptive processing approach and provides the greatest design flexibility. Cross-adaptive effects can be further enhanced using a feedback loop.

In principle, cross-adaptive digital audio effects have been in use since the development of the automatic microphone mixer [24] (see Chapter 1). Automixers balance multiple sound sources based on each source's level. They reduce the strength of a microphone's audio signal when it is not in use, often taking into account the relative level of each microphone.

Verfaille and Zölzer [89] provided brief descriptions of several auto-adaptive effects that could easily be transported into the cross-adaptive world, but their only cross-adaptive suggestion was dynamic time-warping. In [58, 91], auto-adaptive effects from [89] were expanded to cross-adaptive applications. The cross-adaptive architecture includes many potential designs that can be used in full mixing systems, which will be discussed in later chapters.

2.3 Traditional Audio Effects

The use of effects in post-production of music and other art forms can be both technical and creative. Automation is easier for the technical applications since then there may be clear, objective targets which may be obtained without human intervention. If human operators are still involved, they can then focus on the creative aspects of music making.

For this reason, this section focuses on effects which have some objective basis or largely agreed upon best practices. As nearly every device which converts sound to another sound in real time can be considered an audio effect, the range of available processors is vast. Some other typical effect categories include modulation effects (such as chorus, flanger, phaser, vibrato and tremolo), pitch altering, and delays or echoes.

2.3.1 Balance

Balance is widely considered to be the foundation of mixing [6], so much so that mix engineers used to be called 'balance engineers.'

Volume setting in a mix is typically performed by a volume fader, though prior to this there may be also a gain applied to boost the signal to a reasonable range. The fader is marked in decibel units. A typical range may have a minimum value of $-\infty$ dB and maximum value of $+12$ dB. At 0 dB, the fader leaves the volume unchanged.

Determination of the appropriate volume level for each individual track is perhaps the most fundamental mixing task. There are lower and upper bounds, often set by the recording and playback equipment: combined levels should not be so low that the noise floor becomes significant, nor so high that distortion or clipping may occur. Choice of level is also intimately related to perceived loudness, and hence to the frequency content of each track. The task may be quite challenging since perceived loudness of a track is influenced by the presence of other sources. That is, the appropriate relative loudness of tracks A and B will change when track C is introduced.

2.3.2 Stereo Positioning

Stereo audio files encode two signals, or channels, for listening over headphones or loudspeakers, so that sources can be localized.

For reproduction of spatial audio via multiple loudspeakers, we should first consider how we localize sound sources. Consider a listener hearing the same content coming from two different locations, and at slightly different times and levels. Under the right conditions, this will be perceived as a single sound source, but emanating from a location between the two original locations. This fundamental aspect of sound perception is a key element in many spatial audio rendering techniques.

The most standard way to play back two channel stereo sound is over loudspeakers or headphones. The sound sources are typically positioned using level differences (panorama) or sometimes time differences (precedence) between the two channels.

Suppose we have two loudspeakers in different locations. Then the apparent position of a source can be changed just by giving the same source signal to both loudspeakers, but at different relative levels, known as panning. During mixing, this panning is often accomplished separately for each sound source, giving a panorama of virtual source positions in the space spanned by the loudspeakers.

The precedence effect is a well-known phenomenon that plays a large part in our perception of source location. In a standard stereo loudspeaker layout, the perceived azimuth angle of a monophonic source that is fed to both loudspeakers can be changed by having a small time difference between the two channels. If the time difference between the signals fed to the two loudspeakers is below the echo threshold, the listener will perceive a single auditory event. This threshold can vary widely depending on the signal, but ranges from about 5 to 40 ms. When this time difference is below about 1 ms (i.e., much lower than the echo threshold), the source angle can be perceived as being between the two loudspeakers. In between 1 ms and the echo threshold, the sound appears to come just from whichever loudspeaker has the least delay. More generally, the precedence effect is the phenomenon whereby two sounds, when they are apart by an appropriate delay, are perceived as a single entity coming from the direction of the first to arrive.

This effect has several implications. In a stereo loudspeaker layout, if the listener is far enough away from the central position, they will locate the source as being located at the closest loudspeaker and this apparent position does not change even if the other channel is significantly louder.

Subjective testing has shown that an almost equivalent relationship can exist between the effect of time difference and the effect of level difference on perceived azimuth angle. However, the actual relationship can depend on the characteristics of the sounds that are played, the exact placement of the loudspeakers, and the listener [94].

2.3.3 Equalization

Equalization or EQ is one of the most common audio effects. It is the process of adjusting the relative strength of different frequency bands within a signal. The name 'equalization' comes from the desire to obtain a flat (equal) frequency response from an audio system by compensating for non-ideal equipment or room acoustics. Peaks or troughs in the frequency response of a system are often described as 'coloration' of the sound, and equalization can be used to remove this coloration. The term is more generally used for any application of frequency-dependent emphasis or attenuation through filtering.

Equalization covers a broad class of effects, ranging from simple tone controls to sophisticated multiband equalizers. All EQ effects are based on filters, and most equalizers consist of multiple subfilters, each of which affects a particular frequency band. Their use in the music production process has become well-established, with standardized designs for graphic and parametric equalizers, and widely accepted operations to achieve desired tasks [95]. But in recent years there has been an emergence of adaptive, intelligent or autonomous audio equalizers, which provide novel approaches to their design and use.

Equalization, as described in [18], is an integral part of the music production workflow, with applications in live sound engineering, recording, music production, and mastering, in which multiple frequency-dependent gains are imposed upon an audio signal. Equalizing a sound mix is one of the most complex tasks in music mixing and requires significant expertise. The main problem of determining the amount of equalization to be used is that the perceived equalization is different from the applied equalization. In order to achieve a perceptually pleasant equalization several things should be considered: whether or not the channel needs equalization at all, how many filters should be used, the type of filters and ultimately the amount of boost or cut they should have. Some studies on how the sound engineer performs these decisions have been made by Bitzer et al. [92, 96].

Generally, the process of equalization can be categorized as either corrective equalization, in which problematic frequencies are often attenuated to prevent issues such as acoustic feedback, hum [97], masking [98], and salient frequencies [92, 96], or creative equalization, in which the audio spectrum is modified to achieve a desired timbral aesthetic [99] and highlight or hide certain quality of the source. While the former is primarily based on adapting the effect parameters to the changes in the audio signal, the latter often involves a process of translation between a perceived timbral adjective such as *bright*, *flat*, or *sibilant* and an audio effect input space. As music production is an inherently technical process, this mapping procedure is not necessarily trivial, and further complicated by the source-dependent nature of the task.

A parametric equalizer is usually comprised of several peaking filters to increase or reduce gain at and around a certain frequency. These are often complemented by high and low shelving filters which attenuate or boost all frequencies below (low shelf) or above (high shelf) a chosen cut-off frequency by a fixed amount. The user can typically control the center frequency, Q or bandwidth, and gain of the so-called 'bell curve', and the gain, cut-off frequencies and sometimes steepness of the shelving filters.

Parametric equalizers allow the operator to add peaks or troughs at arbitrary locations in the audio spectrum. Adding a peak can be useful to help an instrument be heard in a complex mix, or to deliberately add coloration to an instrument's sound by boosting or reducing a particular frequency range. The ability to cut a frequency can be used to attenuate unwanted sounds, including removing power line hum (50 Hz or 60 Hz and their harmonics) and reducing feedback. To remove artifacts without affecting the rest of the sound, a narrow bandwidth would be used. To modify the overall sound of an instrument by reducing a particular frequency range, a wider bandwidth might be used.

Some other filters aim to remove some region while leaving another unaffected, including the low-pass filter, high-pass filter, band-pass filter, and band stop or notch filter. These can be used to constrain the bandwidth of the source and reduce the spectral noise outside the useful bandwidth. For example, in the case of a piccolo it might be desirable to attenuate undesired noise in the low frequencies using a high-pass filter (also known as a low cut filter), as any energy here can only be due to electronic or physical noise sources.

Inexperienced mixers tend to boost the equalizer parameters more than to cut them. This requires compensating the overall gain of the channel so that the gain remains nominal, and makes it difficult to achieve a stable system with good acoustic gain. This is a tedious and iterative problem, which can be automated.

2.3.4 Compression

As the name implies, dynamic range compression is concerned with mapping the dynamic range of an audio signal to a smaller range. Traditional dynamic range compressors achieve this goal by reducing the high signal levels while leaving the quieter parts untreated, in a process called downward compression. The same effect would be attained by upward compression, where quiet parts are amplified and loud parts are left alone, but this is a matter of (historical) convention. Compression sends the signal along two different paths, one that goes through the analysis circuit (the side-chain), and another that goes through the processing circuit (the voltage controlled amplifier or VCA). While the side-chain signal is often identical to the signal being processed, it can also be a filtered version or even an entirely different signal (external-adaptive).

Considering the classic audio effects (equalization, delay, panning, etc.), the dynamic range compressor is perhaps the most complex one, with a wide variety of applications. It is used extensively in audio recording, noise reduction, broadcasting and live performance. But it does need to be used with care, since its overuse or use with nonideal parameter settings can introduce distortion, 'pumping' and 'breathing' artifacts, and an overly reduced dynamic range, reducing subjective sound quality.

Compressors provide the ability to increase the loudness of a source by bringing up everything but short peaks in level (of albums as well as entire radio channels). Increased loudness can make a source comparatively more appealing, but has led to the much-debated 'loudness war' [100]. In recent years, loudness standards for broadcasting and loudness equalization by streaming services have reduced the competitive advantage of 'mixing loud,' prompting some to say the war is now over.

When recording onto magnetic tape, distortion can occur at high signal levels. So compressors and limiters (a fast-acting compressor with a high gain reduction ratio) are used to prevent sudden transient sounds causing the distortion, while still ensuring that the signal level is kept above the background noise. In fact, professional and consumer noise reduction systems for tape rely on this mechanism, performing a more or less inverse process, expansion, on the playback side.

When editing and mixing tracks, compression can correct for issues that arose in the recording process. For example, a singer may move towards and away from a microphone, so a small amount of compression could be applied to reduce the resultant volume changes. Once tracks have been recorded, a compressor provides a means to adjust the dynamic range of the track. In some cases, compression may be used as an alternative to equalization, altering the source's timbre and taming resonances or louder notes.

Compression may also be used creatively, altering content in more audible ways. There is no single correct form or implementation, and there are a bewildering variety of design choices, each with their distinct sonic character. This explains why every compressor in common usage behaves and sounds slightly different and why certain compressor models have become audio engineers' favorites for certain types of signals.

A popular use of dynamic range compression is to increase the sustain of a musical instrument. By compressing a signal, the output level is made more constant by turning down the louder portions. For example, after a note is played, the envelope will decay towards silence. A compressor with appropriate settings will slow this decay, and hence preserve the instrument's sound. Alternatively, a transient modifier could be used [101], attenuating or boosting just the transient portions (as opposed to the sustain), which include the attack stage of a musical note.

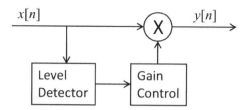

Figure 2.4 Block diagram of a level-dependent, feedforward gain control

2.3.5 Expanders and Noise Gates

The expander is a dynamic processor that attenuates the low-level portions of a signal and leaves the rest unaffected – or indeed amplifying the louder bits instead. As such, it is the converse of a dynamic range compressor (though it cannot entirely undo or invert the compression [102]), and can be viewed as a variable gain amplifier. When the signal level is high, the expander has unity gain. But when the signal drops below a predetermined level, gain will decrease, making the signal even lower. The basic structure of an expander is as a level-dependent, feedforward gain control, identical to the feedforward compressor block diagram (see Figure 2.4).

An important parameter of the expander is the threshold, which is generally user-adjustable. When the input signal level is above the threshold nothing happens, and the output equals the input. But a gain reduction results whenever the level drops or stays below the threshold. This gain reduction lowers the input level, thus expanding the dynamic range. For a typical expander, the signal levels are taken from an average level based on a root-mean-square (RMS) calculation, rather than from instantaneous measurements, since even a signal with high average level may produce low instantaneous measurements.

As with compression, the amount of expansion that is applied is usually expressed as a ratio between the change in input level (expressed in dB) and the corresponding change in output level.

If an expander is used with extreme settings where the input/output characteristic becomes almost vertical below the threshold e.g., expansion ratio larger than 1:8, it is known as a noise gate or simply 'gate.' In this case, the signal may be very strongly attenuated or even completely cut when quiet, and the noise gate will act like an on/off switch for an audio signal. As the name suggests, noise gates are mainly used to reduce the level of noise in a signal.

A noise gate has five main parameters: threshold, attack, release, hold, and gain [42]. Threshold and gain are measured in decibels, and attack, release, and hold are measured in milliseconds. The threshold is the level above which the signal will open the gate and below which it will not. The gain is the attenuation applied to the signal when the gate is closed. The attack and release parameters are time constants representing how quickly the gate opens and closes. Note that this is different from in the compressor, where attack relates to the time to reduce the gain and release to the time to restore to the original levels. The hold parameter defines the minimum time for which the gate must remain open. It prevents the gate from switching between states too quickly which can cause modulation artifacts.

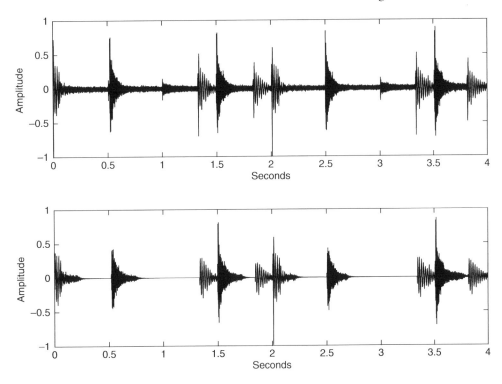

Figure 2.5 (top) Audio recorded from a snare drum microphone, with spill from the cymbals and kick drum in a drum kit; (bottom) the same recording, but with a noise gate applied to remove the low-level bleed from other sources.

There are many audio applications of noise gates. For example, noise gates are used to remove breathing from vocal tracks, hum from distorted guitars, and bleed on drum tracks, particularly snare and kick drum tracks.

Figure 2.5 (top) shows an example snare drum recording containing bleed from secondary sources, and Figure 2.5 (bottom) shows the same signal after a noise gate has been applied. The amplitude of the spill from other sources is very low and will have minimal effect on the gate settings. Components of the bleed signal which coincide with the snare cannot be removed by the gate, because the gate is opened by the snare.

2.3.6 Time Alignment

When recording with a microphone it is important to choose the correct microphone for the task. Different makes, models and types of microphones have different frequency responses, causing spectral changes to the input source, and different polar patterns, which determine the directivity. The choice can be creative, such as choosing a particular microphone because of its peak at the vocal frequencies. It can also be a technical decision to use the microphone that will cause the smallest spectral change to the input signal.

Figure 2.6 A single source reproduced by a single microphone (left) and by two microphones (right). The path of the direct sound is indicated by a dashed line.

The next task is microphone placement, which again is both a creative and technical decision. The microphone may be placed to reproduce certain characteristics of a sound source but the placement can also cause other, unwanted artifacts to occur on the output of the microphone. The simplest configuration is a single microphone placed to reproduce a single source as shown on the left in Figure 2.6. The source is placed within the microphone's pickup area, defined by its polar pattern. The microphone picks up the source, but there may also be interference from reflections, exhibited as reverberation, or noise from other sources [103].

A musical instrument will often reproduce very different sounds from different parts of the instrument. For example, the sound of an acoustic guitar from a microphone placed next to the sound hole will be different to the sound from a microphone near the neck. For this reason, multiple microphones can be used to reproduce different aspects of an instrument. These are then mixed to produce the desired sound for that instrument. An example of this configuration can be seen on the right in Figure 2.6.

Use of more than one microphone may introduce other artifacts. It is difficult, and often not desired, to place multiple microphones equidistant from the sound source. When the distances from each microphone to the source differ, the delay times for a sound to travel to each microphone will also differ. If these microphone signals are summed, either to create a desired sound for the source or mixed to a stereo output, the difference in delay time will cause the resulting output to be affected by comb filtering.

Comb filtering occurs when a signal is summed with a delayed version of itself, such as in the multiple microphone configuration. Cancellation and reinforcement of frequencies occurs periodically between the two signals causing a comb-shaped frequency response, as shown in Figure 2.7. The resultant sound can be described as *thin* and *phasey* and is the basis of the flanger effect [88]. A perceptual study in [104] showed that comb filtering can be heard when the delayed signal is as much as 18 dB lower than the original.

Any configuration that will result in delayed versions of the same signal being mixed can cause comb filtering. For instance, there is an expected latency between a direct-input signal recording an electric guitar and using both a direct-input signal and a microphone recording of its amplifier. In the case of parallel processing, where a track is duplicated and one or both

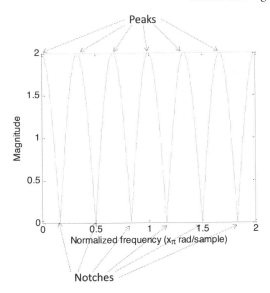

Figure 2.7 The magnitude response of a comb filter with a six sample delay. The locations of the six peaks and six notches cover the whole frequency range.

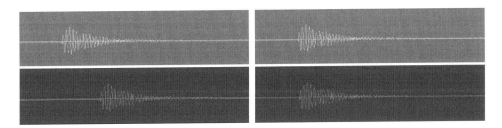

Figure 2.8 Nudging audio regions in an audio editor to compensate for delays that cause comb filtering showing delayed (left) and compensated (right) signals.

versions are processed before being mixed together, comb filtering may occur if there is a difference in latency in the processing applied to either track.

A difference of as little as one sample can cause audible comb filtering to the source signal. At a sample rate of 44.1 kHz and taking the speed of sound at 343 m/s, a difference in time of one sample is equal to a difference in distance of just 8 mm. Summing a signal delayed by one sample with the original signal creates a simple low-pass filter and therefore high frequencies will be attenuated.

Comb filtering can be reduced by applying a compensating delay to give the illusion that the source arrived at both microphones at the same time. This can be done by measuring the distances of the microphones to the source and calculating the difference in delays that occur and applying an additional delay to the microphone signal that has initially the least delay. Compensating delay can also be applied 'by ear' until the comb filtering is reduced. In a studio situation where the audio can be post-processed, the audio regions can be nudged so the signals visually line up. This is shown on the left in Figure 2.8 for a snare drum recorded by two microphones. The bottom waveform contains the delayed signal. The right of Figure 2.8

shows the same recording after the waveforms have been manually 'nudged' into alignment. Many audio production software tools include some form of delay compensation to account for latencies that occur when using effect plugins as inserts.

These methods can be inaccurate and do not attempt to apply the precise delay that is occurring. They also apply a static delay. Therefore if the source or the microphones are moving, comb filtering may still occur.

Though comb filters are manifested as equally spaced notches in the spectrum, they are primarily a time domain problem. This might open the door to some automatic delay correction. For example, when capturing a single source with two microphones that are separated from each other and then mixed together, there will be a comb-filter effect, which will suppress frequencies whose wavelengths are integer multiples of half the distance of the separation of the microphones. Adding the right delay to one of these microphones will avoid this type of artifact. A line input delay in a channel is not a new concept, but setting it up automatically to reduce comb-filtering is.

2.3.7 Reverb

Reverberation (reverb) is one of the most often used effects in audio production. In a room, or any acoustic environment, there is a direct path from any sound source to a listener, but sound waves also take indirect paths by reflecting off the walls, the ceiling or objects before they arrive at the listener, as shown in Figure 2.9. These reflected sound waves travel a longer distance than the direct sound and are partly absorbed by the surfaces, so they take longer to arrive, are weaker than the direct sound, and their frequency components are affected differently across the spectrum. The sound waves can also reflect off of multiple surfaces before they arrive at the listener. The sum of these delayed, attenuated and filtered copies of the original sound is known as reverberation, and it is essential to the perception of spaciousness in the sounds.

Reverberation is not perceived as a mere series of echoes. In a typical room, there are many, many reflections, and the rate at which the reflections arrive changes over time. These reflections are not perceived as distinct from the sound source. Instead, we perceive the effect of the combination of all the reflections.

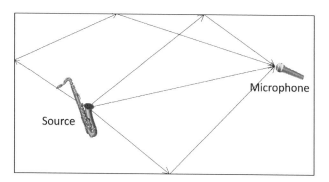

Figure 2.9 Reverb is the result of sound waves traveling many different paths from a source to a listener

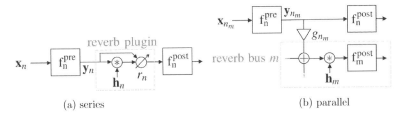

Figure 2.10 Reverb signal chains (from [82])

We usually inhabit the reverberant field, with many sources of reverberation already around us. Yet it is still useful to embed reverb in recordings, to emulate or complement the acoustics of an indoor space [105]. We often listen to music in environments with very little or poor reverb. A dry signal may sound unnatural, so the addition of reverb to recordings is used to compensate for the fact that we cannot always listen to music in well-designed acoustic environments. The reverberation in a car may not adequately recreate the majestic sound of a symphony orchestra. And when listening over headphones, there is no space to add reverberation to the music.

On the other hand, sources are often recorded in spaces which have relatively little reverberation. Generating reverb artificially helps avoid the cost of recording on location. It allows for control over the reverberation type and amount in post-production, compensating for close microphone positions – maximizing isolation from other sources or creatively zooming in on part of a source – or creating unnatural effects that are not constrained by the laws of physics [106, 107]. Early examples of artificial reverberation were generated by re-recording the signal through a speaker in an echo chamber, or an improvised space such as a bathroom or staircase, or emulated using electromechanical devices such as a spring or plate reverb. Contemporary reverberation effects are mostly implemented digitally [107–109].

In mixing consoles and audio workstations, reverberation is typically applied in one of two ways (see Figure 2.10). It can be used as an insert effect on a single track, in which case a wet/dry ratio r_n determines the relative amount of reverb; or several tracks can be sent to a dedicated reverb bus, which runs in parallel to the other tracks and subgroups, and contains only the reverb component of the sum of these tracks.

2.3.8 Distortion

Distortion has always been an important effect in music production, and is a crucial characteristic of many analog audio devices including microphones, amplifiers, mixing consoles, effect processors and analog tape. Distortion is often a nuisance in the context of electronics design, but a subtle, pleasant sounding distortion is one of the most sought-after features in audio equipment, even (or especially) after the advent of digital audio production [87].

Here, the term distortion is used to denote the application of a nonlinear input-output characteristic (sometimes referred to as transfer function) to a waveform, a process typically referred to as harmonic distortion, amplitude distortion or waveshaping, as opposed to phase or time-domain distortion [107, 110–112]. In the digital domain, this can be any nonlinear

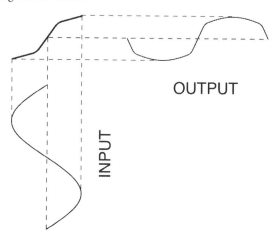

Figure 2.11 Example of amplitude distortion applied to a sine wave

function defined, for simplicity, from $[-1, 1]$ (input) to $[-1, 1]$ (output) (see for example Figure 2.11). As such, it is a 'low-cost' audio effect which can be implemented through calculation of a simple mathematical function or a lookup table [110].

Distortion as a creative, often dramatic effect owes its popularity to rock guitarists, but it can alter timbre and dynamics in sophisticated and subtle ways for a large variety of sources [113]. Beyond its creative, sound-shaping use, harmonic distortion can be of practical importance by increasing the audibility of sources [63] or as an alternative for equalization [107]. In contrast to EQ, it can introduce frequency components which are not present in the original signal, for instance to brighten a dull recording. Harmonics generated by a distortion effect can further create the illusion of a low fundamental, even if this fundamental is later filtered out or inaudible on certain playback systems. By increasing the loudness for the same peak level, a soft clipping curve can be a suitable substitute for fast-acting dynamic range compression. Carefully applied distortion can even enhance the perceived quality and presence of audio signals [113].

3

Understanding the Mix

"Like a book, a recording reduces the listener to the position of passive consumer."

O. B. Hardison, Jr., *Disappearing Through the Skylight*,
Penguin, New York, 1989.

3.1 Fundamentals and Concepts

3.1.1 Terminology

The production chain for recorded and live music, from conception to consumption, consists of several stages of creative and technical processes. Compositions or improvisations materialize as acoustic vibrations, which are then captured, sculpted and rendered. Figure 3.1 shows a simplified depiction of such a music production chain.

Between the performance of the music and the commitment of audio signals to the intended medium, the different sources are transformed and merged into one consolidated signal, in a process known as the mix. The mastering engineer usually picks up where the mixing engineer left off. Mastering involves, among other things, adjusting and signing off on the loudness, timbre, dynamic profile, and trailing silence of a song, both on its own and as part of an album [115].

The distinction between mixing and mastering herein is blurry, as both use similar tools to attain similar goals. The mix engineer indirectly affects the overall spectral contour and dynamic range by processing each source, and may even apply compression and equalization to the mix bus. Excessive use of bus compression is frowned upon as it treads on the mastering engineer's domain and irreversibly reduces the dynamic range. Conversely, mastering engineers often receive separate stems in addition to a consolidated stereo (or

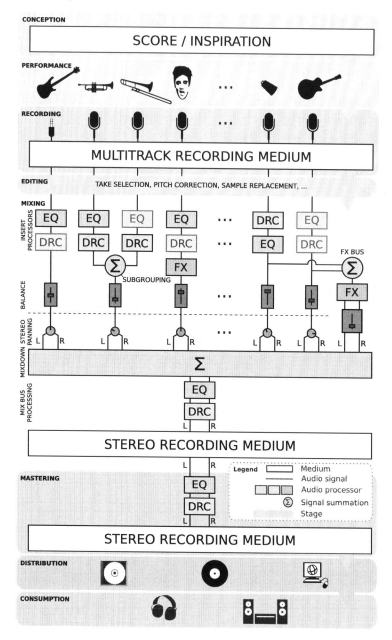

Figure 3.1 Example of a music production chain in the case of a stereo studio recording (from [114])

multichannel) mix, so they may still apply some processing of different instrument groups individually [116].

In the context of this research, remixing and mixing is not the same. Mixing refers to combining recorded audio signals and spectrally or dynamically modifying them. On the other hand, remixing refers to editing audio sources, some of which have been previously mixed, and splicing, looping, reordering and scaling them in time to create a substantially different piece of music. For example, a DJ might change the tempo of a mixed song to extend its length and match a previous song. The final remix is a longer, more complex variation of the original. Some confusion about these terms exist, especially as remixing occasionally denotes 'mixing again.' Remastering is also used rather liberally for any kind of processing – be it restorative or creative – of old material for a new release, not just typical audio mastering processes.

3.1.2 The Role of the Mix Engineer

Mixing music is itself a complex task that includes dynamically adjusting levels, stereo positions, filters, dynamic range processing parameters and effect settings of multiple audio streams [108]. Mix engineers are expected to solve technical issues, such as ensuring the audibility of sources, as well as to make creative choices to implement the musical vision of the artist, producer, or themselves [117]. As such, they occupy a space between artist and scientist, relying on creativity as well as specific domain knowledge [207].

The mix starts after the recording phase, based on the resulting tracks, which the mix engineer is then tasked to process and combine into a near final version of a song. The process involves correcting artifacts that have arisen from the recording process (the fix), processing the tracks so that they fit nicely and blend well together (the fit), and highlighting tracks or aspects of the multitrack audio in order to heighten the artistic impact of the final mix (the feature) [118].

3.1.3 The Mixing Workflow

An assumption that is often made about mixing is that it is an iterative process [108, 119]. But how is this iteration performed? Izhaki [108] suggests that a coarse-to-fine approach is applied whereby mixing decisions are continually refined using a somewhat standard ordering of audio effects. This is supported by Jillings and Stables [120], who measured decreasing magnitude of adjustments over the course of a mix project. In contrast, Ma [121] views mixing as an optimization problem, where a particular aspect is mostly perfected before moving on to the next aspect, with targets and criteria set for the final mix. These two iterative approaches are depicted in Figure 3.2. Both views lend themselves well to an intelligent systems approach, whereby the steps can be sequenced and diverse optimization or adaptive techniques can be applied in order to achieve given objectives.

Different engineers have different approaches to where to start a mix, and in which order to proceed. Pestana [119] states that there is no fixed order in the sequence of steps applied. Popular choices include having all faders down or all faders up, and begin by processing either the drums (bottom-up) or the lead vocal (top-down) [122]. Similarly, two main approaches exist for mixing drums: using overhead microphones as the main signal and adding emphasis as needed with the close kick and snare drum microphones, or using the close microphones as primary signals and bringing up the more distant microphones for added 'air' or 'ambiance' to taste [55]. Though these considerations may seem arbitrary, research has shown that the starting position of faders has significant impact on the final result [120]. Moreover, engineers starting with the same initial balance will tend to arrive at a similar outcome, even if their trajectories are very different [123].

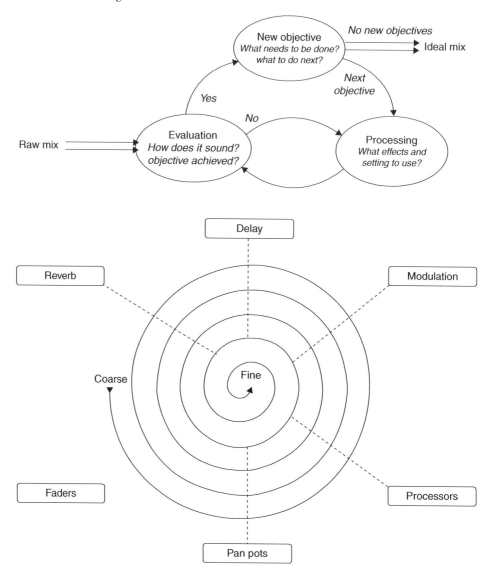

Figure 3.2 The iterative approaches to mixing multitrack audio, as described by [121] (top) and [108] (bottom)

3.1.4 The Stage Metaphor

A popular representation of the mix process is to adjust the positions of the sources in a virtual live venue, to the extent that some digital mixing interfaces are presented as a stage with draggable instrument icons (see Section 9.2.1). In this scenario, the mix engineer is crafting the listener's illusion of attending an imaginary concert, with the instruments positioned appropriately. An equally important mental image is the perspective of a musician: a piano's lows are commonly panned left and highs panned right, and drum sets are laid out over the scene according to their physical position in a (right-handed) drum set. As a result, many engineers have strong opinions on whether they should adopt a 'drummer perspective' or an 'audience perspective.'

This positioning is achieved through simultaneous use of just about every tool available to the mix engineer. Reverb is often quoted as the primary device for adding the dimension of 'depth' to a mix. Width is achieved mostly by panning or by using 'stereo' sources where the constituent channels are subtly different and widely spaced. Frequency or pitch perceptually correlates with height [124, 125]. Perceived distance can further be affected by adjusting level (closer sources are louder), equalization (a faraway source has attenuated high frequencies due to air absorption, and a very nearby source has increased low frequencies due to the proximity effect) and reverberation (a higher level and shorter predelay move the source further away). While the recording medium poses few constraints on the perceived locations of instruments, it is still often inspired by typical stage setups, and mostly abides by the rules of physics.

Conversely, actual stage positions in live performance may be based on sonic considerations. In rock and pop music, it is customary to see the drums and lead vocalist in the center, and guitars, keyboards and brass sections balancing each other on either side. In large venues, the effect is merely visual, but can be inspired by the typical positions in a mix. In more intimate settings, where a large portion of the sound comes from the stage, the physical position of drums and guitar amplifiers have a genuine impact on the sound and the audience's ability to spatially resolve different instruments. Even when the PA system overpowers the 'acoustic' sound coming from the stage, the precedence effect can still result in a source's perceived position to match its position on the stage, if the original sound arrives at the listener just before the amplified version does [126].

3.1.5 Order of Processes

The mixing setup is usually very modular. Processors can be inserted in an almost limitless number of configurations, especially in a digital environment where few elements are 'hard-wired.' Research about the suggested order of signal processing operations is ongoing, and most practical literature bases the preferred sequence on workflow considerations [108, 127]. However, when effects are nonlinear, the resulting sound will be different depending on their relative position. For instance, a dynamic range compressor will affect the signal differently if an EQ is placed before instead of after it [128]. An EQ stage is often desired before the compressor, because a heavy low end or a salient frequency may trigger the compressor when it is not desired [108, 118, 129].

Faders and pan pots generally manipulate the signal after the effects processors such as compressor and equalizer. Mixing consoles are generally wired with pan pots after faders. But because of the linear nature of these these two processes and their independence in this system, the sequencing is of no importance in this context.

3.2 Mix Assumptions

It may not be possible to compile a single set of rules underpinning the esoteric process of mixing. However, some mixes are clearly favored over others, suggesting that there are 'best practices' in music production [130]. Many audio engineering handbooks report standard settings for mixing for various instruments, genres and desired effects. Some of these 'rules' are contradictory and very few have been validated [54]. The same sources containing such mixing rules also state that mixing is highly nonlinear [118] and unpredictable [129], and that there are no hard and fast rules to follow [118], 'magic' settings [107] or even effective equalizer presets [129].

The typical spectral and dynamic processing of tracks depends very much on the characteristics of the input signal [9], but presets provide a starting point for many novice and professional sound engineers. Essentially 'deaf' to the incoming signal, these presets are instead based on assumptions about the signal's properties, and are sometimes intended for a particular instrument.

Important questions arise concerning the psychoacoustics of mixing multitrack content. For instance, little has been formally established concerning user preference for relative amounts of dynamic range compression used on each track. Admittedly, such choices are often artistic decisions, but there are many technical tasks in the production process for which listening tests have not yet been performed to even establish whether a listener preference exists.

Listening tests must be performed to ascertain the extent to which listeners can detect undesired artifacts that commonly occur in the audio production process. Important work in this area has addressed issues such as level balance preference [131,132], reverberation level preference [133, 134], 'punch' [135], perceived loudness and dynamic range compression [136], as well as the design and interpretation of such listening tests.

As the perception of any one source is influenced by the sonic characteristics of other simultaneously playing sources, the problem of mixing is multidimensional. Consequently, the various types of processing on the individual elements cannot be studied in isolation only. There are several important considerations that affect the direction of research in this field and help to establish the functionality of Intelligent Music Production tools. These include decisions concerning

- which aspects of the production process may be performed by an intelligent system;
- whether or not the goal is to mimic human decisions;
- whether the system should operate in real time or render a mix offline;
- whether the mix is static or time-varying; and
- whether the mix will be played back in a particular acoustic space.

Different assumptions apply to the processes of mixing for the different situations. And different challenges are addressed using different methods when designing intelligent mixing systems for them.

3.2.1 *Live Versus Pre-recorded*

Live mixing differs from mixing in the studio in several aspects, the most obvious of which is its ephemeral character: the sound is processed in real time, as the musicians are playing and as the audience is listening. The live sound engineer also knows the acoustics and playback system through which their mix will be consumed. However, the listener position varies wildly, and optimizing the sound for those in the 'sweet spot' only is not acceptable.

Panning between left and right speakers is largely pointless, as this would benefit only audience members who are equidistant from both speakers (or at the same relative position as the sound engineer), with everyone else essentially hearing the loudspeaker closest to them.

To complicate things, sound comes directly from the stage and can impose a significant lower limit to a source's loudness – especially in the presence of loud instruments like acoustic drums. Acoustic feedback (the 'screeching' or 'howling' noise that arises when a microphone picks up its own signal through a speaker, resulting in an unstable system) also needs to be taken into account. Feedback determines the upper limit to the amplification factor of the various acoustic instruments, along with the maximum sound pressure level permitted or desired in the interest of auditory safety. As a result, even the mundane objective of making the vocal as loud as the drum set can be a considerable challenge.

A related task is monitor mixing. Distinct from front of house mixing for the audience, the goal here is to produce customized mixes for each musician to hear those instruments they would like to hear, including their own. Aesthetics are secondary to clarity and audibility, but the challenges of avoiding feedback while overpowering 'noise' (such as loud drums or audience cheers) are still present.

Mixing live for broadcast is another special case. The transient aspect of the mix process remains, but the content may be experienced through a wide range of systems in different acoustics, immediately or at a later point in time.

In the case of studio mixing, the overall playback level is dictated by the medium and any source can be arbitrarily loud relative to another. The result should neither distort nor have a relatively audible noise floor. When sources are isolated from each other, or even recorded separately, the 'spill' or acoustic interference of other instruments in a microphone recording is no longer a challenge. Mixing pre-recorded content means that there is rarely a requirement for real-time processing. The operator – human or machine – can take the entirety of the recorded content into account when deciding how to process any given instant. Any 'mistakes' can be mitigated after the fact, and there is usually the opportunity to apply extensive creative enhancements. As a result, there are more possibilities for intelligent systems in the realm of studio mixing, but also a higher expectation in terms of quality and creativity.

3.2.2 Venue Acoustics

For a small room, there is rarely a need to amplify high sound pressure level instruments like trombones. Thus the mix inside the mixing console might be completely different from the acoustic mix heard by the audience, and may contain only those acoustic sources with a low sound pressure level.

In a large room or open space, all sources may be reinforced. The sources do not interact with the acoustics of the room, thus removing a challenging complication. In these situations, we can often approximate the mix happening inside the mixing console to be similar to the delivered acoustic mix. We can thus take advantage of some of the assumptions made when dealing with pre-recorded tracks. For this reason, much of the existing Intelligent Music Production research has mainly focused on post-production with pre-recorded tracks, or on large rooms for live production tasks.

3.2.3 Static Versus Time Varying Mix

In sound production, time-varying mix parameters are implemented through automation. This is currently applied using either scenes or automation tracks. The scene approach uses snapshot representations of a mix, which are recalled at certain points in time. This facilitates changing the state of the mixing parameters dynamically. There is usually some sort of interpolation mechanism between snapshots to reduce any artifacts introduced when recalling a scene. Similarly, automation tracks are the equivalent of a timeline representation of the state of a parameter and are common in digital audio workstations (DAW).

In recorded media, the mixing of a single song is often comprised of several automation scenes, while for live performances the tendency is to have a smaller number of scenes. In general, for live mixing, a static mix is built first, usually during the soundcheck, and then the mixing engineer enhances it manually during the show. In current systems these scenes are prerecorded. This means the programming of a scene requires prior knowledge of the input signal and must be reprogrammed if the input sources are to be changed.

Due to the complexity of dynamic parameters, most prior research on sound production has been restricted to time invariant mixes, even though the systems are adaptive and capable of real-time use.

3.3 Challenges

3.3.1 Cultural Influence

There is evidence that the sound of recorded and mixed music can be influenced by the studio's location and the engineer's background, to a point where its origin might be reliably determined solely based on sonic properties [127, 137, 138]. The same sources suggest that these differences have been disappearing due to the increased mobility of information, people and equipment. But the findings of most of those studies are based on subjects from a single type of background, so their wider ecological relevance can be questioned.

However, concurrence of the findings further strengthens the support for the initial conclusions. Such was the case with [139, 140], which both highlighted cultural and generational differences in mixing practices. They identified a stronger tendency to 'mix in the box' among younger engineers, and differences in average mix loudness for mixing students from different schools. And on a larger scale, [141, 142] found that definitions of sound-related adjectives vary between different countries.

3.3.2 Effect of Experience

The consistency of a mix engineer, as well as the agreement between different engineers, has been shown to be considerably larger with highly trained individuals. For instance, more experienced engineers show less variance over time when they are repeatedly asked to set the level of a solo instrument accompanied by a background track [131]. More experienced engineers are also more critical and specific in their evaluation of mixes by others, and more likely to agree with each other on these assessments than amateur music producers [139].

3.3.3 Genre

Supposed rules underpinning the mixing process cannot be generalized across genres of music. For instance, many assumptions are specific to instruments common in pop music (lead vocal, kick drum, bass guitar), but are not applicable if those instruments are not present or if they have a different role in the composition.

The mixing paradigm is inherently different for most classical and some acoustic and traditional music, where the aim of the recording or amplification process is to recreate the experience of a hypothetical audience member attending an acoustic performance (realism) or a subtly enhanced version thereof (hyperrealism) [143]. To this end, many decisions are made prior to recording, by committing to a particular space, microphone placement, and ensemble layout. The post-recording process then consists of balancing and panning the various microphone feeds, and only subtle dynamic range and temporal processing. As the recorded tracks often contain a high level of crosstalk, the timbre of single instruments cannot be extensively altered. Furthermore, as faithful imitation of the sound stage requires knowledge of the instrument and microphone positions, the recording and mixing engineer work closely together when they are not the same person.

On the other side of the spectrum is electronic, dub and experimental music. Here, timbres are shaped by the performer for creative effect. In the case of synthesis, parts are often designed to have a particular frequency spectrum and dynamic profile. Even recorded sources like vocals are often heavily processed to fit the artistic vision of the performer, including reverberation, delays, filters and distortion. In some cases, the mixing console and effects are used as instruments, contributing to the performance as much as any instrumental part. A dedicated mix engineer can still improve the result, but their actions depend very much on what degree of processing has already been applied. As such, a general description of their method is difficult to produce.

What remain are those musical genres where the mix engineer has a mostly defined role that can be seen as separate from the role of the recording engineer or performer. The task at hand is not to recreate a soundscape in the most realistic way, but to enhance and alter the sounds to meet the audience's expectations. For instance, even the softest vocal is made audible throughout and positioned front and center, and various components of the drum set are spread across the stereo spectrum – cymbals wide and spacious, and kick drum close and tight.

Of course, this division is not rigid. Many examples of modern music include electronic and acoustic elements, and some electronic music features traditional pop music templates including a bass instrument, kick and snare drum, and vocal.

3.3.4 Approaches and Subjectivity

A final obstacle in the creation of intelligent music production systems is the wide variety of mixing approaches and styles preferred by producers or consumers [144]. This strengthens the argument for customization of otherwise automatic tools, so that a template can be chosen to correspond to the desired aesthetic or paradigm. At the same time, it is a common and justified criticism of some IMP tools, the results of which are very limited in variety, where professional engineers would achieve a rich diversity of outputs. This is therefore an important consideration when tools are meant to produce a technically accurate result, that is also aesthetically pleasing and generates interest.

Part II
How Do We Construct Intelligent Structures?

4

IMP Construction

"Supposing, for instance, that the fundamental relations of pitched sounds in the science of harmony and of musical composition were susceptible of such expression and adaptations, [the Analytical Engine] might compose elaborate and scientific pieces of music of any degree of complexity or extent."

Augusta Ada King-Noel, Countess of Lovelace, notes on L. F. Menabrea's "Sketch of the Analytical Engine Invented by Charles Babbage," *Scientific Memoirs*, 1843.

4.1 Intelligent and Adaptive Digital Audio Effects

Rather than have sound engineers manually apply many audio effects to all audio inputs and determine their appropriate parameter settings, intelligent and adaptive digital audio effects may be used instead [89]. The parameter settings of adaptive effects are determined by analysis of the audio content, achieved by a feature extraction component built into the effect. Intelligent audio effects also analyze or 'listen' to the audio signal, are imbued with knowledge of their intended use, and control their own operation in a manner similar to manual operation by a trained engineer. The knowledge of their use may be derived from established best practices in sound engineering, psychoacoustic studies that provide understanding of human preference for audio processing techniques, or machine learning from training data based on previous use. Thus, an intelligent audio effect may be used to set the appropriate equalization, automate the parameters on dynamics processors, and adjust stereo recordings to more effectively distinguish the sources, among many other things.

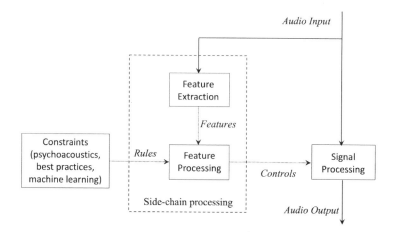

Figure 4.1 Block diagram of an intelligent audio effect. Features are extracted by analysis of the audio signal. These features are then processed based on a set of rules intended to mimic the behavior of a trained engineer.

An example of an intelligent audio effect is shown in Figure 4.1. Here, the side-chain is essential for low latency, real-time signal processing flow. The audio signal flow remains unaffected, but any feature extraction and analysis is performed in parallel in a separate section. In [80], Reiss describes several of these intelligent, adaptive effects, for use with single channel audio. They automate many parameters and enable a higher level of audio manipulation. These intelligent audio effects include adaptive effects that control the panning of a sound source between two user defined points depending on the sound level or frequency content of the source, and noise gates with parameters which are automatically derived from the signal content.

4.2 Building Blocks of Intelligent Audio Effects

4.2.1 Feature Extraction

An audio feature is a characteristic scalar, vector or matrix, measured from a signal. Some features may have little or no correlation with human perception, such as peak level, bit depth and sample rate. These are often referred to as low-level features, and can come from electronic constraints or technical limitations of an audio device. Other low-level features, such as spectral centroid or zero crossing rate, come directly from statistical properties of the acoustic signal. High-level features are those that provide a symbolic, abstract, or semantic representation of a signal, and are often correlated with human perception. These often form the basis of classifiers such as genre or musical key detectors, and are typically engineered to work well at a given task. Mid-level representations, such as pitch and onsets, bridge the gap by incorporating some basic but still musically relevant features.

The feature extraction block of an intelligent audio effect is in charge of extracting a series of features per input channel. The ability to extract the features quickly and accurately will determine the ability of the system to perform appropriately in real time. The better the feature extraction model, the better the algorithm will perform. For example, if the audio effect aims to address some aspect of perceived loudness, the loudness model chosen to extract the feature will have a direct impact on the performance of the system. According to their feature usage, automatic mixing tools can be accumulative or dynamic.

Accumulative IMP tools store a converging data value which improves in accuracy proportionally to the amount and distribution of data received. They have no need to continuously update the data control stream, which means that they can operate on systems that are performing real-time signal processing operations, even if the feature extraction process can be non-real-time. The main idea behind accumulative automatic mixing tools, is to obtain the probability mass function [145] of the feature under study and use the most probable solution as the driving feature of the system. That is, we derive the mode, which corresponds to the peak value of the probability density of the accumulated extracted feature.

Dynamic IMP tools make use of features that can be extracted quickly in order to drive data control processing parameters in real time. An example of such a system can be one that uses an RMS feature to ride vocals against background music. Another example can be gain sharing algorithms for controlling microphones such as originally implemented by Dugan in [24]. Such tools do not tend to converge to a static value.

A trade-off between dynamic and accumulative feature extraction can be achieved by using relatively small accumulative windows with weighted averages. Processing high-level features in real time is often desirable for improving the perceptual accuracy of intelligent effects, however due to computational load this is often not viable. Studies such as Ward et al. [146] explore methods of supporting this process by applying optimization to a number of bottle-necks in a loudness model. This involves processes such as changing model parameters such as hop-size, and using more efficient filtering techniques.

4.2.2 Feature Extraction with Noise

The reliability of feature extraction methods is influenced by the existence of bleed, crosstalk, ambient noise or other phenomena that cause the source signal to be polluted with unwanted artefacts. Common methods for obtaining more reliable features include averaging, gating and pre- or post-processing. For example, a popular method used for automatic microphone mixing is adaptive gating, where the gating threshold for the microphone signal $x_m[n]$ adapts to the existing noise [24, 26]. This requires an input noise source $x_n[n]$ that is representative of the noise present in the system. For a live system, a microphone outside of the input source capture area could provide a good representation of ambient noise. This reference can then be used to derive the adaptive threshold needed to validate the feature, as in Figure 4.2.

Although automatic gating is generally used to change the dynamic characteristics of an audio signal, it can also be implemented on the data features extracted from the signal as opposed to directly applied to the signal. This has the advantage of being less processing intensive.

For accumulative tools, variance threshold measures can be used to validate the accuracy of the probability mass function peak value. This effectively measures the deviation of the signal from its average, rather than the average itself. The choice of feature extraction model will influence the convergence times in order to achieve the desired variance. For this to work well in real time in a system that is receiving an unknown input signal, some rescaling operations must be undertaken. If the maximum dynamic range of the feature is unknown,

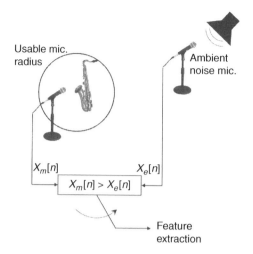

Figure 4.2 Diagram of an adaptive gated system

then the probability mass function must be rescaled. In such a case, the axis range should be continuously normalized to unity by dividing all received feature magnitudes by the magnitude of the maximum received input value.

4.2.3 Side-chain Processing

Due to the complexity of the signal processing involved in most expert systems, it would seem unlikely that true real-time processing could be achieved. But real-time signal processing flow with low latency is required for live performance applications. Thus side-chain processing can be performed where the audio signal flow happens in a normal manner in the signal processing device, such as a digital mixer, while the required analysis, such as feature extraction and classification of the running signal, is performed in a separate analysis instance. Once the required amount of certainty is achieved on the analysis side, a control signal can be sent to the signal processing side in order to trigger the desired parameter control change command.

A dynamic range compressor is typically implemented whereby the level of the signal is computed in the side-chain that produces a gain value, which is used to process the input audio signal. For standard feedforward designs, the same audio signal is used as input to the side-chain. However, if the side-chain takes another signal as input, it becomes an external-adaptive effect.

Side-chaining is used extensively in modern music production, where it often refers more specifically to an external-adaptive effect that uses envelope following as the analysis algorithm. Many hardware and software compressors have a 'side-chain' input. This is intended for an auxiliary audio input, as opposed to the default input signal, to be used as a control signal for the amount of compression to apply. Such side-chain compression may be used to have a pumping effect in dance music. For instance, by routing the kick drum to the side-chain of a compressor and a sum of other tracks to the compressor's main input,

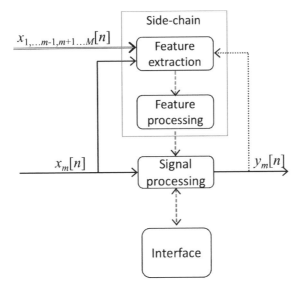

Figure 4.3 The mth track of a cross-adaptive processing device with multitrack input, with optional feedback (dotted line)

the volume of the other tracks can rapidly attenuate every time the kick drum is hit and then swell back.

Side-chaining systems which use an external input are common in many software plugins, and side-chain routing is available in most popular DAWs. While the effect is predominantly used to control a compression parameter using the amplitude envelope of another signal, side-chaining is an adaptive technique that has gained widespread use and is a good indicator as to the potential for external-adaptive effects that use more sophisticated analysis algorithms.

4.3 Cross-adaptive Processing

The architecture for intelligent audio effects described so far assumes that each effect processes a single audio track. At most, they may extract features from an additional input in the side-chain. But when mixing multitrack audio, one performs signal processing changes on a given source not only because of the source content but also depending on the properties of many other sources. For instance, it may be deemed necessary to blend a given source with the content of other sources, so that a high-quality mix is achieved. That means the relationships between many sources may have to be taken into account.

This can be conceptualized as an extension of the cross-adaptive audio effect model, as described in Section 2.2.3, in which the signal processing of an individual source is the result of analysis of the features of each source and the relationships between all involved sources. Such an effect makes use of a Multiple Input Multiple Output (MIMO) architecture, and may be considered inter-dependent across channels.

This is depicted in Figure 4.3 for one track within a multitrack, cross-adaptive audio effect. Assuming the cross-adaptive tool has the same number of inputs and outputs, inputs may be given as $x_m[n]$ and outputs as $y_m[n]$, where m has a valid range from 1 to M given that M is the maximum number of input tracks involved in the signal processing section of the tool.

A cross-adaptive multitrack audio effect is typically comprised of two main sections, the signal processing section and the side-chain processing section. The signal processing algorithm may be a standard audio effect and can include a user interface if the tool is meant to provide visual feedback or metering for its actions. The side-chain consists of a feature extraction block, and a cross-adaptive feature-processing block.

The feature extraction block will extract features from all input tracks. Consider the feature vector $\vec{f}_m[n]$ obtained from the feature extraction section for source m. This may correspond to different features in each cross-adaptive audio effect. For example, in time alignment it may correspond to a temporal value extracted using cross-correlation, whereas in spectral enhancement the feature vector may consist of values taken from applying a spectral decomposition to the signal.

These feature values are then sent to the cross-adaptive feature processing block. Feature processing can be implemented by a mathematical function that maps the relationships between the input features to control values applied in the signal processing section. Typically, either the feature values or the obtained control values are smoothed before being sent to the next stage. This can be achieved using an exponential moving average filter.

As with the feature vector, the control vector $\vec{c}_m[n]$ will correspond to different parameters according to the aim of each multitrack audio effect. For example, suppose the objective is to achieve equal level for all channels, where $l_m[n]$ is the level per track, and the processing per track is given by $y_m[n] = c_m[n] \cdot x_m[n]$. A simple control could be given by $c_m[n] = l_{avg}[n]/l_m[n]$, which ensures equal level for all channels.

4.4 Mixing Systems

In studies by Reiss and Perez Gonazalez [80, 147], and references therein, several cross-adaptive digital audio effects are described, which explore the possibility of reproducing the mixing decisions of a skilled audio engineer with minimal or no human interaction. Each of these effects produces a set of mixes, where each output may be given by applying the controls in the following manner:

$$mix_l[n] = \sum_{m=1}^{M} \sum_{k=1}^{K} c_{k,m,l}[n] * x_m[n], \tag{4.1}$$

where there are M input tracks and L channels in the output mix. K is the length of the control vector c and x is the multitrack input. Thus, the resultant mixed signal at time n is a sum over all input channels, of control vectors convolved with the input signal.

In fact, any cross-adaptive digital audio effect that employs linear filters may be described in this manner. For automatic faders and source enhancement, the control vectors are simple scalars, and hence the convolution operation becomes multiplication. For polarity correction, a binary valued scalar, ± 1, is used. For automatic panners, two mixes are created, where panning is also determined with a scalar multiplication (typically, employing the sine-cosine panning law). For delay correction, the control vectors become a single delay operation. This applies even when different delay estimation methods are used, or when there are multiple active sources [148]. If multitrack convolution reverb is applied, then c represents direct application of a finite room impulse response. Automatic equalization employs impulse responses for the control vectors based on transfer functions representing each equalization curve applied to each channel. And though dynamic range compression is a nonlinear effect due to its level dependence, the application of feedforward compression is still as a simple gain function. So multitrack dynamic range compression would be based on a time varying gain for each control vector.

In [149], a source separation technique was described where the control vectors are impulse responses that represent IIR unmixing filters for a convolutive mix. Thus, each of the resultant output signals mix_l in Equation 4.1 represents a separated source, dependent on filtering of all input channels.

The approach taken in [41] attempted to deliver a live monitor mixing system that was as close as possible to a predefined target. It approximated the cause and effect relationship between inputs to monitor loudspeakers and intelligibility of sources at performer locations. The stage sound was modeled as a linear MIMO system which enabled all performer requirements to be considered simultaneously. Simple attenuation rules from the monitors to the performers were used in order to perform simulations in a free field environment. The target mix was defined in terms of relative sound pressure levels (SPLs) for each source at each performer. Thus, a constrained optimization approach was used to generate scalar valued control vectors that resulted in optimized mixes at selected positions on stage.

The reverse engineering of a mix, described in [150], assumes that Equation 4.1 holds when presented with original tracks and a final mix. It then uses least squares approximation to estimate each control vector as a fixed length FIR filter. By assuming that the gain in the filter represents the faders, the initial zero coefficients represent delay, differences between left and right output channels are based on sine-cosine panning, and that anything remaining represents equalization, it can then reverse engineer the settings used for time-varying faders, panning, delays and equalization. However, this would not be considered an intelligent audio effect since it requires little or no knowledge of preferred mixing decisions.

4.5 Automatic Mixing Systems

Due to the importance of the relationships between sources when mixing music, we can formulate a design objective to be performed by many IMP systems. This is that the signal processing of an individual source should depend on the relationships between all involved sources. This objective could be met by the use of cross-adaptive processing. Using this as context, Figure 4.4 demonstrates the architecture of an intelligent multitrack mixing system. Here, the side-chain consists of a set of feature extraction modules, applied to each track, and a single analysis section that processes the features extracted from many tracks. The cross-adaptive processors exploit the relationships between input features in order to output the appropriate control data. This data controls the parameters in the signal processing of the multitrack content. This cross-adaptive feature processing can be implemented by a set of constrained rules that consider the interdependence between channels.

The side-chain processing section performs the analysis and decision making. It takes audio from one or more tracks together with optional external inputs, and outputs the derived control data. The controlling data drives the control parameters back to the signal processing algorithm, thus establishing a cycle of feature extraction analysis and decision making. It is characteristic of adaptive effects and therefore characteristic of IMP tools.

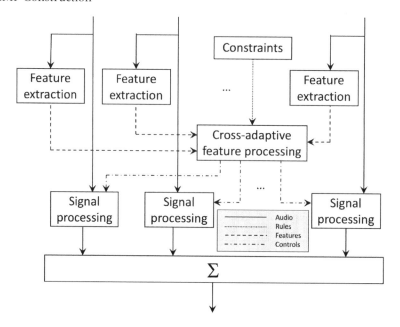

Figure 4.4 Block diagram of an intelligent, cross-adaptive mixing system. Extracted features from all channels are sent to the same feature processing block, where controls are produced. The output channels are summed to produce a mix that depends on the relationships between all input channels. This is presented in more detail by Perez Gonzalez [151].

This notion of adapting parameters in real time can lead to results which go beyond the typical automation of mix parameters (usually track level) and apply the classic processors in entirely new ways, such as dynamically carving out frequency content of a backing track in response to the lead vocal's features [12].

4.6 Building Blocks of Automatic Mixing Tools

We emphasize the generation of an acceptable autonomous mix, which can later be enhanced by a human operator if desired. The proposed approach differs from current approaches in that it requires minimal or in most cases no prior knowledge of the inputs. It can also be adapted to different input without the need for extensive human intervention.

The system takes an autonomous approach to mixing by reducing user interaction and taking into account minimal or no aesthetic considerations and is meant to work in real time for the purpose of live mixing of music. The proposed system is envisaged as a helper to the sound engineer and composer, rather than giving the engineer an additional set of constraint parameters to be manipulated. The approach seeks to enhance the user experience by automatically mixing the input sources while reducing or eliminating the technical mixing tasks required to be performed by the user. This allows the engineer to concentrate on the aesthetic aspects and performing the fine-tuning involved in generating a pleasing sound mix.

Although the system mainly deals with the technical constraints involved in generating an audio mix, the developed system takes advantage of common practices performed by sound engineers whenever possible. The system also makes use of inter-dependent channel information for controlling signal processing tasks while aiming to maintain system stability at all times. Working implementations are described and comparative subjective evaluation of mixes has been used to measure the success of IMP tools.

Automatic mixing explores the possibility of reproducing the mixing decisions of a skilled audio engineer with minimal or no human interaction. Several independent systems are presented that, when combined together, can generate an automatic sound mix out of an unknown set of multichannel inputs. In an automatic mixing context we often seek to aid or replace the tasks normally performed by the user. Our aim is to emulate the user's control parameters. In order to achieve this some important design objectives should be performed by the tools:

1. The system should comply with all the technical constraints of a mix, such as maintaining adequate dynamic range while avoiding distortion.
2. The design should simplify complex mixing tasks while performing at a standard similar to that of an expert user.
3. For sound reinforcement applications, such as live music or live performance inside an acoustic environment, the system must remain free of undesired acoustic feedback artefacts.

Low-level features such as RMS, peak levels and bit depth characterize the technical limitation of electronic devices. Such features are often used in standard signal processing, effects devices, and effects embedded in mixing consoles. But for mixing musical signals, the sound engineer attempts to make the right balance of technical and perceptual attributes. That is, one often attempts to achieve a perceptual goal, using mostly tools governed by technical requirements. Given that we have a reliable set of perceptual features extracted from the input signal, a simple yet elegant solution for achieving a balance between both the technical and perceptual space can be achieved by average normalization. This process is shown in Figure 4.5.

When average normalization is used with a perceptual feature it can be used to balance the ratio of the perceptual feature. Such a method can also be used with low-level features to maintain technical requirements. For instance, when used with a feature such as gain it can keep a system under stability, therefore avoiding unwanted acoustic feedback artefacts.

Many IMP tools aim to make objective technical decisions. This is useful for improving the audio engineer's workflow and allowing them to achieve a well-balanced mix in a shorter period of time. For this, common mixing practices can often be used as constraints. Given that the task normally performed by an expert user also involves perceptual considerations, perceptual rules can improve the performance of the algorithms. When combining several basic automatic mixing tools to emulate the signal path of a standard mixer, we can achieve a mix in which the signal processing flow is comparable to the one performed in a standard mixing situation. A set of mixing tools, which make extensive use of the concepts herein will be presented. Some of these tools are more exploratory, and designed to take into account uncommon mixing practices or to be able to take subjective mixing decisions.

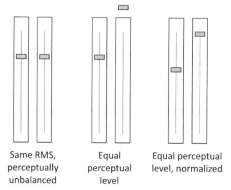

Same RMS, perceptually unbalanced	Equal perceptual level	Equal perceptual level, normalized

Figure 4.5 On the left is an example of fader settings that might result in an equal low-level feature (RMS), but unequal perceptual feature (loudness). The middle shows settings producing equal loudness for the two channels but is technically impossible. On the right is a perceptually balanced and technically possible solution using average normalization (explored in [151]).

4.6.1 Reference Signals and Adaptive Thresholds

An important consideration to be taken into account during analysis of an audio signal is the presence of noise. The existence of interference, crosstalk and ambient noise will influence the ability to derive information about the source. For many tasks, the signal analysis should only be based on signal content when the source is active, and the presence of significant noise can make this difficult to identify.

One of the most common methods used for ensuring that an intelligent tool can operate with widely varying input data is adaptive gating, where a gating threshold adapts according to the presence of noise. A reference microphone placed far from the source signal may be used to capture an estimation of ambient noise. This microphone signal can then be used to derive the adaptive threshold. Although automatic gating is typically applied to gate an audio signal it can also be used to gate whether the extracted features will be processed.

The most straightforward way to implement this is to apply a gate that ensures the control vector is only updated when the signal level of the mth channel is larger than the level of the reference, as given in the following equation:

$$c_m[n+1] = \begin{cases} c_m[n] & x_{RMS,m}[n] \le r_{RMS}[n]) \\ \alpha c'_m[n+1] + (1-\alpha)c_m[n] & otherwise \end{cases} \tag{4.2}$$

where c'_m represents an instantaneous estimation of the control vector. Thus the current control vector is a weighted sum of the previous control vector and some function of the extracted features. Initially, computation of RMS level of a signal x is given by

$$x_{RMS}[n] = \sqrt{\frac{1}{K}\sum_{k=0}^{K-1} x^2[n-k]} \tag{4.3}$$

Later values may either be given using a sliding window, which reduces to:

$$x_{RMS}[n+1] = \sqrt{x^2[n+1]/K + x^2_{RMS}[n] - x^2[n+1-K]/K} \tag{4.4}$$

or a one pole low-pass filter (also known as an exponential moving average filter):

$$x_{RMS}[n+1] = \sqrt{\beta x^2[n+1] + (1-\beta)x_{RMS}^2[n]} \qquad (4.5)$$

Here, the values represent time constants of IIR filters, and allow for the control vector and RMS estimation, respectively, to smoothly change with varying conditions. Equation 4.4 represents a form of dynamic real-time extraction of a feature (in this case, RMS), and Equation 4.5 represents an accumulative form.

4.6.2 System Stability

For live performance mixing, an automatic system must avoid undesired artefacts caused by acoustic feedback. For this reason several stability solutions have been developed, for example gain sharing [24, 26] and self-normalization techniques. In most cases these techniques try to prevent acoustic feedback by ensuring a maximum electronic transfer function gain no larger than unity. This ensures that regardless of the changes in signal processing parameters the system remains stable.

4.6.3 Live Processing

The standard approach adopted by the research community for real-time audio signal processing is to perform a direct translation of a computationally efficient offline routine into one that operates on a frame-by-frame basis. Effective use in live sound or interactive audio however, requires not only that the methods be real-time, but also that there is no perceptible latency. The minimal latency requirement is necessary because there should be no perceptible delay between when a sound is produced and when the modified sound is heard by the listener. Thus, many common real-time technologies, such as look-ahead and the use of long windows, are not possible. The windowed approach produces an inherent delay – the length of a window – that renders such techniques impractical for many applications. Nor can one assume temporal invariance, as sources move and content changes during performance. To surmount these barriers, perceptually relevant features must be found which can be quickly extracted in the time domain, analysis must rapidly adapt to varying conditions and constraints, and effects must be produced in advance of a change in signal content.

4.7 Intelligence

4.7.1 Automated, Automatic, Autonomous

It is important to distinguish between automatic mixing processes and automated mixing processes. Parameter automation is a widely used production technique and is a feature that can be found in many modern mixing consoles and DAWs. Parameter automation is the recording of manual adjustments that are made to effects parameters throughout the duration of a track. DAWs achieve this by recording an automation curve that is then displayed below or overlaid on the track that the audio effect is applied to. Once recorded, the automation curve may be visually adjusted for synchronization to musical events. The end result of parameter automation is thus similar to an adaptive effect, with the difference being that a human, rather than an algorithm, is doing the musical analysis. This technique could be extended by providing the option of using an analysis algorithm for initial recording of the automation curve.

An automatic process involves autonomous actions, i.e., systems or processes that can perform desired tasks in unstructured environments without continuous human guidance. This autonomous process can be treated as a constrained rule problem in which the design of the control rules determines the process to be applied to the input signals. Currently, commercial automatic mixing does not take advantage of the full recalling capabilities of automated mixers and is only capable of gain control.

Here, it is instructive to consider a definition of a fully autonomous robot. It must have the ability to gain information about the environment, work for an extended period without human intervention, and move either all or part of itself throughout its operating environment without human assistance [152]. An autonomous robot may also learn or gain new capabilities like adjusting strategies for accomplishing its task(s) or adapting to changing surroundings.

4.7.2 Incorporating Best Practices into Constrained Control Rules

In order to develop intelligent software tools, it is essential to formalize and analyze audio production methods and techniques. This will establish the required functionality of such tools. Furthermore, analysis of the mixing and mastering process will identify techniques that facilitate the mixing of multitracks, and repetitive tasks which can be automated. By establishing methodologies of audio production used by professional sound engineers, features and constraints can be specified that will enable automation. Here, we describe three different approaches to doing this, based on formal approaches to laying the groundwork for an expert system.

Knowledge Engineering

Knowledge engineering is an approach to intelligent or expert system design that seeks to integrate human knowledge into computer systems in order to solve challenging problems that normally would require a high level of human expertise [153]. As a discipline, it is closely related to cognitive sciences, since the knowledge may be organized and utilized according to our understanding of human reasoning.

The process of constructing a knowledge-based system is often loosely structured as follows:

1. Assessment of the problem
2. Development of a knowledge-based system shell
3. Acquisition and structuring of the related information, knowledge and specific preferences
4. Implementation of the structured knowledge into knowledge bases
5. Testing and validation of the inserted knowledge
6. Integration and maintenance of the system
7. Revision and evaluation of the system.

For intelligent audio production we know that the knowledge lies in the hands of top practitioners, but extracting it is not trivial. This is partly because practical sound engineering has moved away from a technical to an artistic field in the last half a century, and partly because practitioners are often inclined to believe there is no knowledge implied. This mirrors well-known challenges in knowledge engineering for many domains of application.

There are two main views of knowledge engineering. In the traditional transfer view, human knowledge is transferred into a knowledge base, and then into artificial intelligence systems. An underlying assumption is that the human knowledge is available and usable. But typically, people, including (or even especially) experts, have implicit knowledge that is essential to problem solving.

However, the required knowledge may not be captured in a knowledge base. This has led to the modeling view, where one attempts to model the problem solving techniques of domain experts in addition to the explicit knowledge. But since this view needs to capture a dynamic and iterative process, it may not utilize the most efficient and direct use of the captured knowledge.

The application of structured methods for knowledge acquisition and organization can increase the efficiency of the acquisition process, but one should not ignore the wide variety of relevant data and methods. Each type of knowledge may require its own approach and technique. Methods should be chosen to reflect the different types of experts and expertise. Many ways of representing the acquired knowledge might be applied, since this can aid the acquisition, validation and reuse of knowledge. And the knowledge can be used in several ways, so that the acquisition process can be guided by the project aims.

An important aspect of many knowledge engineering systems is the use of role limiting methods, which employ various knowledge roles where the knowledge expected for each role is clarified. Configurable role limiting methods assume that a problem solving method can be divided into smaller tasks, where each task can be solved by its own problem solving method. These generic tasks typically each have a rigid knowledge structure, a standard strategy to solve problems, a specific input and a specific output.

A major concern in knowledge engineering is the construction of ontologies, which define the terms inside a domain and the relationships between those terms. Ontologies for audio production have been devised, and will be discussed further in Section 5.3.2.

One principle often (but not always!) applied in knowledge engineering is the strong interaction problem hypothesis. This states that the structure and representation of domain knowledge is completely determined by its use. This implies that ontologies and knowledge bases are not easily transferable from one problem to the next.

Many of the best practices in sound engineering are well-known, and have been described in the literature [12, 54]. In live sound for instance, the maximum acoustic gain of the lead vocalist, if present, tends to be the reference to which the rest of the channels are mixed, and this maximum acoustic gain is constrained by the level at which acoustic feedback occurs. Furthermore, resonances and background hum should be removed from individual sources before mixing, all active sources should be heard, delays should be set so as to prevent comb filtering, dynamic range compression should reduce drastic changes in loudness of one source as compared to the rest of the mix, panning should be balanced, spectral and psychoacoustic masking of sources must be minimized, and so on.

Similarly, many aspects of sound spatialization obey standard rules. For instance, sources with similar frequency content should be placed far apart, in order to prevent spatial masking and improve the intelligibility of content. A stereo mix should be balanced and hard panning is avoided. When spatial audio is rendered with height, low-frequency sound sources are typically placed near the ground, and high frequency sources are placed above, in accordance with human auditory preference. Interestingly, Wakefield et al. [154] show that this avoidance of spatial masking may be a far more effective way to address general masking issues in a mix than alternative approaches using equalizers, compressors and level balancing. Also, many parameters on digital audio effects can be set based on analysis of the signal content, e.g., attack and release on dynamics processors are kept short for percussive sounds. These best practices and common approaches translate directly into constraints that are built into intelligent software tools.

For example, Perez Gonzalez and Reiss [38] measure the spectral masking in a mix to enhance a source while minimizing the changes in levels of the other sources, similar to the way in which a sound engineer attempts to make only slight changes to those sources that make a mix sound 'muddy.' This is achieved by using filterbanks to find the dominant frequency range of each input signal. Channels with closely related frequency content are highly attenuated while minimizing the attenuation of those channels with frequency content far from the source, e.g., if one wants to enhance the stand-up bass in a mix, the kick drum will be attenuated but there will be little change to the flute. This method also uses knowledge of sound engineering practice since its operation is similar to that of a parametric equalizer, commonly used for boosting or attenuating a frequency range in single channel audio processing.

Scott and Kim [55] describe a mixing system focused on drum tracks. It applies common practices based on instrument types. These practices were found mainly by interviews with professional and student mixing engineers, and by recommended practices described in [108, 129]. For example, close microphones were used as primary sources and overhead microphones used to increase the amount of cymbals and add 'air' to the mix, and kick and snare drums levels were higher than overhead and toms. In terms of stereo panning, kick and snare were placed at central positions, toms were spaced linearly left to right, and overhead tracks panned alternating left and right. Equalization decisions were made to give the kick drum more punch and the snare more brilliance.

A knowledge engineering approach to music production was broadly applied by De Man and Reiss [54, 155]. In these studies, the authors describe autonomous systems that are built entirely on best practices found in audio engineering literature. Also, many parameters on digital audio effects can be set based on an understanding of best practices and analysis of signal content, e.g., attack and release on dynamics processors are kept short for percussive sounds.

Grounded Theory

Grounded theory is the systematic generation of theory from data that contains both inductive and deductive thinking. Informally, one can think of it as the reverse of the scientific method. Rather than state a hypothesis which is then confirmed by experiment, an experiment is performed which leads to formulating hypotheses that may be explained by a theory. One grounded theory approach is to formulate hypotheses based on conceptual ideas, but a more rigorous approach is to verify the hypotheses that are generated by constantly comparing conceptualized data on different levels of abstraction, and these comparisons contain deductive steps. Another goal of a grounded theory study may also be used to simply uncover the hypotheses that seem most important or relevant.

Though grounded theory is well-established in the social sciences, it is used frequently throughout research fields, and often without the approach being formalized. The use of pilot studies, leading to proposing a hypothesis that may then be tested in a larger formal study, contains many grounded theory ideas. In fact, whenever data is being gathered and researchers ask "What is happening in the experiment, what is the interesting behavior, how and why are complex phenomena or behavior arising?," then they are applying the type of inductive reasoning necessary to form a theory grounded in empirical data.

In the behavioral sciences, grounded theory has the goal of generating concepts that explain how people resolve their central concerns regardless of time and place. The use of description in a theory generated by the grounded theory method is mainly to illustrate concepts. If the goal is simply accurate description, then another method should be chosen.

The unit of analysis is the incident, and in many behavioral experiments, there may be many incidents for each participant. When comparing incidents, the emerging concepts and their relationships are in reality probability statements. Consequently, grounded theory is a general method that can use any kind of data even though the most common use is with qualitative data. However, although working with probabilities, most grounded theory studies are considered as qualitative since statistical methods are not used, and figures are not presented. The results are not a reporting of statistically significant probabilities but a set of probability statements about the relationship between concepts, or an integrated set of conceptual hypotheses developed from empirical data.

Glaser [156] proposed that a grounded theory should not be judged on its validity (though of course, that is an essential criterion for a mature theory), but instead should be judged by the following four concepts:

- *Fit* has to do with how closely concepts fit with the incidents they are representing, and this is related to how thoroughly the constant comparison of incidents to concepts was done.
- *Relevance* is concerned with whether the study deals with the real world concerns, captures the attention and is not only of academic interest.
- *Workability* is the extent to which the theory explains how the problem is being solved with much variation.
- *Modifiability* is whether the theory can be altered when new relevant data is compared to existing data.

Grounded theory has, so far, not been exploited much in the development of IMP systems. A wealth of studies in the social sciences have tried to extract meaning from interviews and studies of music producers and sound engineers, but this rarely generalizes to common themes. However, some formal approaches utilize subjective studies to formulate rules and hypotheses. For audio production, this translates to psychoacoustic studies that define mix attributes, and perceptual audio evaluation to determine listener preference for mix approaches. An important downside of this approach is that it is very resource intensive. Perhaps the most relevant grounded theory approaches can be found in the comprehensive analysis of music production practices by Pestana [12,119], and the focused, formal grounded theory approach in the work of Pras [8, 157–161].

Machine Learning

A third approach to devising IMP systems, and one that is more aligned with current scientific and engineering approaches, is machine learning. This is a vast field, but here we are concerned with systems that can be trained on initial content and infer a means to manipulate new content, without being provided with fixed rules and their parameters. An initial example in the context of autonomous music production was Scott et al. [44], where a machine learning system was provided with both stems and final mixes for 48 songs. This was used as training data, and it was then able to apply time varying gains on each track for new content. However, this approach is limited due to the rarity of available multitracks, and the related copyright issues. This is in stark contrast to music informatics research, usually concerned with the analysis of mixed stereo or monaural recordings, where there is a wealth of available content. More recently with the advent of deep learning, IMP systems that exploit the power of convolutional neural networks have been deployed for solving the problems of automatic mixing and equalization [162, 163].

4.8 Concluding Remarks

At first glance, it may seem that we have now established all the basic building blocks of IMP systems. We know the architecture of intelligent audio effects and mixing systems. We know the components required, and we know how to imbue the systems with intelligence. But we do not know what this intelligence might be. It is not enough to say, for instance, that we can find out by applying machine learning techniques. Where is this data that we can use for machine learning? Multitracks and their mixes are prized commercial content. And simply saying that we can do listening tests to establish rules of mixing is not sufficient. How can we scale this up to large numbers of tests? How do we establish themes and hence hypothesize new rules which may not be in the knowledge base? How do we confirm the proposed rules?

It is clear that two more important aspects must be discussed. First, there need to be ways to gather huge amounts of information about music production. Second, we need to refine and elucidate the perceptual evaluation approaches that are required. These two points motivate the following two chapters, which constitute the remainder of this section.

5

Data Collection and Representation

"I propose that there will be a new branch of professional audio that will arise over the next 20 years, and this is the intelligent assistant. [...] This level of knowledge can be used increasingly to take over the mundane aspects of music production, leaving the creative side to the professionals, where it belongs."

James A. Moorer, "Audio in the new millennium,"
Journal of the Audio Engineering Society, vol. 48,
no. 5, pp. 490–498, 2000.

One of the key challenges in engineering is to gather large sets of relevant, representative data. This data forms the basis of a wide number of tasks, from training and testing models, to the subjective evaluation of empirically developed systems. In the field of Intelligent Music Production, this is particularly important as we tend to focus on the development of systems that rely very heavily on the approximation of subjective decisions. For instance, in order to develop algorithms that mix like an audio engineer, we first need to establish how an audio engineer mixes. In the absence of a complex set of defined rules, one can imagine extracting this information and identifying trends by analyzing many mixes.

In this chapter, we discuss a number of key datasets for IMP and show how they address some of the main challenges in the field. Additionally, we present a number of methods for collecting this information, and demonstrate the complexity of gathering useful, representative data.

5.1 Datasets

A range of useful data stores have been tried and tested in the field of music production. Many of these datasets formed the basis for the intelligent systems presented in the latter sections of this book. They have been essential for model training, testing, and subjective evaluation, and all capture some aspects of the human decision making process at various points in the sound production workflow.

5.1.1 Multitrack Audio

Many types of audio and music research rely on multitrack recordings for analysis, training and testing of models, or demonstration of algorithms, and the field of Intelligent Music Production is no exception. Where training on large datasets is needed, such as with machine learning applications, a large number of audio samples is especially critical. There is no shortage of mono and stereo recordings of single instruments and ensembles, but the study or processing of multitrack audio suffers from a severe lack of relevant material [3]. This limits the generality, relevance, and quality of the research and the designed systems.

An important obstacle to the widespread availability of multitrack audio is copyright. It grants the owner of an original work exclusive rights to determine under what conditions it may be used by others. But it may also restrict the free sharing of most pieces of music and their components [144]. This impedes reproducing or improving on previous studies, limits the extent to which the data can be made public, and hinders comparing against other work. Data that can be shared without limits, on account of a Creative Commons or similar license, facilitates collaboration, reproducibility and demonstration of research and even allows it to be used in commercial settings, making the materials appealing to a larger audience. In addition to research, the availability of this type of data is also useful for budding mix engineers, audio educators, developers, as well as musicians or creative professionals in need of accompanying music or other audio where some tracks can be disabled [164]. Some Creative Commons content has formed the basis of mixing competitions and sound production courses across the world.

Table 5.1 provides a list of the most commonly used multitrack resources at the time of publication, along with an indication as to whether the resource is free to use and distributed under a Creative Commons license. The following paragraphs document notable examples which have been widely used in IMP research.

MedleyDB

Among these datasets, MedleyDB is used extensively for MIR and IMP research, as it features a large number of annotated, royalty-free multitrack recordings [169]. It was initially developed to support research on melody extraction, but is generally applicable to a wide range of multitrack research problems. At this time, the dataset contains over 250 multitracks, from a variety of musical genres but with an emphasis on classical and jazz, which account for about 50% of the content [176]. The data comes from a range of professional musicians, either recorded specifically for the project in recording studios at NYU, or through other audio education organizations such as Weathervane Music. The multitrack recordings are supplied with song-level metadata, which includes details of the artists, composers, genre, and track names, and stem-level metadata which includes instrument tags.

Table 5.1 Table of existing multitrack resources, with estimated number of songs, whether (some) multitracks are available under a Creative Commons license or are in the public domain, and whether they are freely available

Name	# songs	CC	Free	Ref.
BASS-dB	20	✓	✓	[165]
Bass Bible	70			[166]
Converse Rubber Tracks		(✓)	(✓)	1
David Glenn Recording			(✓)	2
Dueling Mixes				3
DSD100	100		✓	[167]
MASS		✓	✓	[168]
MedleyDB	122	✓	(✓)	[169]
MIR-1K	1000		✓	[170]
Mix Evaluation	19	✓	✓	[139]
Mixing Secrets	180		✓	[129]
MixOff.org			✓	4
MIXPLORATION	3			[405]
MUSDB18	150	(✓)	(✓)	[171]
QUASI	11	(✓)	✓	[172]
Rock Band	48		(✓)	[173]
Shaking Through	50	✓		5
Structural Segmentation	104	(✓)	✓	[174]
Telefunken Microphones			✓	6
TRIOS	5	✓	✓	[175]
Ultimate Metal Forum			✓	7

Mix Evaluation Dataset

At this time, the Mix Evaluation Dataset contains around 20 multitrack recordings, most of them freely available [139]. However, for each song, a considerable number of mixes are available – sometimes more than 25 – with the complete DAW session, so that the mix can be recreated and analyzed in detail [74]. In addition to this, extensive comments [177] and numerical preference ratings [178] from subjective evaluation by participants from several countries are associated with each mix. This allows for extensive research into the attributes of music production that are often lost when session information is discarded. By including deep metadata, the authors provide training data for fully automated music production systems.

Source Separation Data

Source separation datasets are inherently useful for music production due to their reliance on measuring original stems against signals taken from decomposed mixes. Because of this

---

1 www.conversesamplelibrary.com
2 davidglennrecording.com
3 www.duelingmixes.com
4 www.mixoff.org
5 https://weathervanemusic.org/shakingthrough
6 https://telefunken-elektroakustik.com/multitracks
7 www.ultimatemetal.com/forum

they lend themselves well for multitrack mixing research. One of the major downsides are that there are often no mixing parameters provided, such as gain and pan, although a number of them include stems before and after audio effects have been applied. The source separation datasets discussed here were part of the Signal Separation Evaluation Campaign (SiSEC) [172], a recurring large-scale competition in which source separation algorithms are compared and bench-marked against a range of objective metrics.

MASS:[8] Music Audio Signal Separation [168] was the sole dataset used for the early SiSEC challenges. It contains stems from several files ranging from 10 to 40 seconds in length, which are provided before and after audio effect plugins have been applied. Where possible, the authors also provide descriptions of the audio effects used, along with other metadata such as lyrics.

MIR-1K:[9] A dataset collected at the Multimedia Information Retrieval lab, National Taiwan University, comprising 1000 song clips for singing voice separation [170, 179]. The samples from the dataset are separated into two channels, where one represents the singing voice and the other represents the accompaniment. Metadata is provided in the form of manual annotations of pitch, phoneme types, and aligned lyrics.

QUASI:[10] Quaero Audio Signals comprises 11 multitracks collected by researchers at the INRIA and Télécom ParisTech institutes as part of the QUAERO Project [172, 180]. The dataset includes decompositions and mixes of commercial tracks such as "Good Soldier" by Nine Inch Nails Some of the data is licensed under Creative Commons, however some have restrictive licenses. The tracks are mixed by an engineer at Télécom ParisTech to meet a number of criteria such as: balance mixing (no effects), panoramic (gains and pan), just EQ, various effects.

DSD100 and MUSDB18:[11] The Demixing Secrets Dataset comprises 100 selected tracks from the Mixing Secrets database, which were used for the 2016 SiSEC Campaign [181], and the MUSDB18 dataset comprises 150 full length music tracks selected from Mixing Secrets and MedleyDB which were used for the 2018 SiSEC Campaign [171]. Both datasets follow the format illustrated in Figure 5.1, in which mixes are structured into hierarchical sub-components which separate the vocals from the accompaniment.

The MUSDB18 data is also stored in Native Instruments' Stems format,[12] a useful multichannel audio format, discussed later in Section 5.3.1.

5.1.2 Open Multitrack Testbed

Existing online resources of multitrack audio content have a relatively low number of songs, show little variation in content, restrict use due to copyright, provide little to no metadata, lack mixed versions including the parameter settings, or do not come with facilities to search the content for specific criteria. To address this need, an open testbed of multitrack material was launched, integrating a variety of shareable contributions (including the two aforementioned datasets) and accompanying metadata necessary for research purposes.

---

8 www.mtg.upf.edu/static/mass/resources
9 https://sites.google.com/site/unvoicedsoundseparation/mir-1k
10 www.tsi.telecom-paristech.fr/aao/en/2012/03/12/quasi
11 https://sigsep.github.io
12 www.native-instruments.com/en/specials/stems

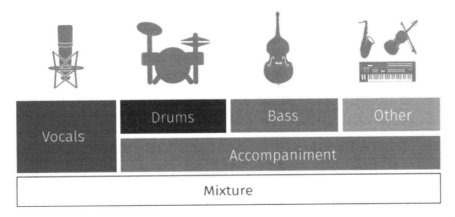

Figure 5.1 Structure of the DSD100 and MUSDB18 datasets, as described by SigSep, 2018 (https://sigsep.github.io).

In order to be useful to the wider research community, the content should be highly diverse in terms of genre, instrumentation, and technical and artistic quality, so that sufficient data is available for most applications. Where training on large sets of data is needed, such as with machine learning applications, a high number of audio samples is especially critical. Furthermore, researchers, journals, conferences, and funding bodies increasingly prefer data to be open, as it facilitates demonstration, reproduction, comparison, and extension of results. A single, widely used, large, and diverse collection of multitracks unencumbered by copyright accomplishes this. Moreover, reliable metadata can serve as a ground truth that is necessary for applications such as instrument identification, where the algorithm's output needs to be compared to the 'actual' instrument. Providing this data makes the testbed an attractive resource for training or testing such algorithms as it obviates the need for manual annotation of the audio, which can be particularly tedious if the number of files becomes large. In addition, for the testbed to be highly usable it is mandatory that the desired type of data can be easily retrieved by filtering or searches pertaining to this metadata. By offering convenient access to a variety of resources, the testbed aims to encourage other researchers and content producers to contribute more material, insofar as licenses or ownership allow it.

For this reason, the Open Multitrack Testbed [315]

- can host a large amount of data;
- supports data of varying type, format, and quality, including raw tracks, stems, mixes, and digital audio workstation (DAW) files;
- contains data under Creative Commons license or similar (including those allowing commercial use);
- offers the possibility to add a wide range of meaningful metadata;
- comes with a semantic database to easily browse, filter, and search based on all metadata fields.

Launched in 2014, the Testbed is continually expanded with locally and remotely hosted content. At the time of writing, it contains close to 600 *songs*, of which some have up to

Title	Artist	Composer	Song Type	Recording Engineer	Number of Tracks	Number of Mixes	Number of Stems
Meet the Frownies	Twin Sister	Twin Sister	studio music	Amy Morrissey	34	3	34
Rise Up Singing	Jackie Greene	Jackie Greene	studio music	David Simon Baker	46	78	0
'La Lyra' Overture	King's College London Baroque Orchestra	Georg Philipp Telemann	studio music	Callum McGee	9	3	0
Night Owl	A Classic Education		studio music	Brian McTear	22	4	20
Waiting for GA	Faces on Film	Faces on Film	studio music	Joe Bisirri	15	4	19
Disturbing Wildlife	Invisible Familiars	Invisible Familiars	studio music	Joe Bisirri	28	22	14
Lush	The Tontons	The Tontons	studio music	Matt Schimelfenig	20	29	8
Save Me	Pattern is Movement	Pattern is Movement	studio music	Matt Schimelfenig	28	9	16
Prisoner's Cinema	The Dead Milkmen	The Dead Milkmen	studio music	Matt Schimelfenig	27	23	27
Stranger	Circuit des Yeux	Circuit des Yeux	studio music	Matt Schimelfenig	41	21	23
Green and Yellow	The Dove and the Wolf	The Dove and the Wolf	studio music	Matt Schimelfenig	26	21	27
Perfect Day	Cassandra Jenkins	Cassandra Jenkins	studio music	Matt Schimelfenig	35	29	15
You Always Look for Someone Lost	Peter Matthew Bauer	Peter Matthew Bauer	studio music	Matt Schimelfenig	30	28	29
New Skin	Torres	Mackenzie Scott	studio music	Matt Schimelfenig	21	13	10

Figure 5.2 Open Multitrack Testbed – Browse interface

300 individual constituent *tracks* from several *takes*, and others up to 400 *mixes* of the same source content. Where licensing allows it, the resources are mirrored within the Testbed. For less liberal or unclear licenses, the metadata is still added to the database, but links point to third party websites.

A wide range of metadata is supported, and included to the extent that it is available for the different items. Using established knowledge representation methods such as the Music Ontology [182] and the Studio Ontology [207] (see Section 5.3.2), *song* attributes include title, artist, license, composer, and recording location; *track* attributes include instrument, microphone, sampling rate, number of channels, and take number; and *mix* attributes include mixing engineer, audio render format, and DAW name and version. These properties can be used to search, filter, and browse the content to find the desired audio.

To quickly find suitable content, the web application includes browse and search functionality (Figures 5.2 and 5.3), to allow filtering and searching using the various metadata properties.

The database offers a SPARQL endpoint to query and insert data through HTTP requests. The infrastructure further supports user accounts and different levels of access, for instance when licenses are less liberal, and a convenient metadata input interface.

With a dataset of this size and diversity, and such a wide range of metadata available, the testbed can be and has been used for various research topics including automatic track labelling [183], source separation and automatic remixing [67], training and testing of intelligent audio effects [61,69], and analysis of music production practices [348]. The Open Multitrack Testbed can be accessed via multitrack.eecs.qmul.ac.uk.

5.1.3 Audio Processing Data

In conjunction with multitrack recordings, other aspects of the music production workflow can be captured specifically, thus providing a basis to develop intelligent and automated

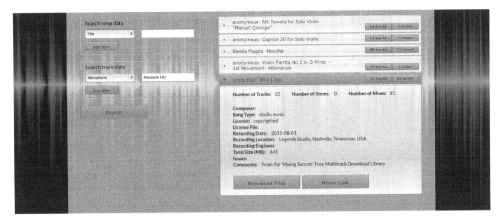

Figure 5.3 Open Multitrack Testbed – Search interface

systems. In addition to mixing, the production process consists of a range of audio processing tasks with varying levels of complexity, often applied using a number of audio effects plugins. In order to build systems that facilitate these tasks, we can utilize user data, whether this it is composed of large-scale usage statistics, or user-labeled production decisions. These types of datasets are generally less commercially sensitive as they tend not to contain raw audio data, and users can be anonymized. The datasets can come in a number of forms, from time-varying parameter manipulation, to static feature sets relating to some aspect of user-defined signal processing.

Descriptive Audio Processing

Tools for semantic or descriptive audio processing allow a user to alter sound by providing human-readable descriptions of some form of signal processing. In order to collect data to support this, researchers gather descriptions or annotations provided by experienced users with associated parameter settings.

The SAFE Dataset,[13] discussed further in Section 5.2.1, consists of audio engineering parameter settings, with matching descriptions and audio features [184]. The dataset was collected in order to specifically understand the relationship between language and audio effects. The data is gathered anonymously through a suite of open-source DAW plugins and, as a result, it is still growing. As of 2016, 2694 submissions had been gathered [81], which represent descriptions of audio effect transformations from four plugin types: a parametric equalizer (1679), a dynamic range compressor (454), an overdrive distortion (303), and an algorithmic reverb (258). The entries include a list of 100 audio features before and after the audio processing is applied, the parameter settings of the plugin, a description of the transform, and a list of additional metadata tags including the genre, instrument, age, and location of the producer.

The SocialFX project[14] encompasses data from a suite of web-based tools for gathering audio descriptions and associated processing [185]. The project amalgamates data from

---

[13] http://semanticaudio.co.uk
[14] http://music.cs.northwestern.edu/research.php

several experiments including 1102 descriptions of graphic equalization (SocialEQ [186]), 2861 of algorithmic reverb (Reverbalize [187]) and 1112 of dynamic range compression. The data was gathered from a wide range of users, with varying levels of music production experience. The descriptions were taken from effects being applied to a predetermined set of sound sources, which are supplied along with parameter settings for each sample.

Recommendation and Detection Data

In addition to semantic control of audio effects, a number of IMP systems are concerned with decomposition or recommendation based on a number of factors taken from the user's environment. For example, we may be able to provide plugin recommendations to a user based on the instrument they are currently processing, or based on the spectral profile of their sound source. We may also want to analyze a processed or mixed signal with a view to reverse engineering the audio processing that has been applied [150]. To do this, the following datasets provide a means to analyze audio processing in music production.

The Processing Chain Recommendation Dataset was gathered using a modified version of the SAFE dataset, which has been adapted for the web [128]. The data captures a range of plugin selections, via an interface which allows users to cascade plugins to process a range of predetermined audio samples. The project gathered 178 processing chains from 47 participants. The effect transforms are distributed with metadata such as the genre and instrument sample, along with a descriptive term provided to the user, based on the original SAFE dataset.

The IDMT-SMT-Audio-Effects[15] data was gathered by the IDMT group at Fraunhofer [188]. The dataset is designed to facilitate the recognition and classification of audio effects applied to musical signals. The dataset contains over 55,000 recordings of bass and guitar notes, with a range of various effects applied to them, including delay, reverb, chorus, tremolo, and distortion. The authors used this data to train a machine learning algorithm with spectral features taken from the dataset to classify according to audio effect type.

5.1.4 Other Notable Datasets

Other than multitrack audio and annotated production data, an extensive range of relevant datasets are available from a wide number fields. In this section, we present a few notable examples which have either been used or are of potential use to IMP research. This section is intended mainly for inspiration. More in-depth repositories of datasets include Alexander Lerch's summary of MIR datasets,[16] or the ISMIR datasets list.[17]

Commercial Recordings

One of the more common types of sound production data is recorded music. This is much more widely available than multitrack audio due to its commercial nature. Though less useful for many aspects of understanding music production, it has formed the basis of a large body

---

[15] www.idmt.fraunhofer.de/en/business_units/m2d/smt/audio_effects.html
[16] www.audiocontentanalysis.org/data-sets
[17] http://ismir.net/resources.html

of genre and instrument classification research, which in turn contribute to a number of semantic mixing studies.

The Million Song Dataset[18] provides audio features for one million commercial audio recordings [189]. The dataset is by far the largest of its kind and has supported hundreds of studies from music recommendation to mood, genre and instrument classification. Because of copyright restrictions it is not distributed with raw audio samples, but each song includes metadata in the form of Artist and Song fields, and EchoNest audio features such as loudness, timbre and tempo.

Similarly, the GTZAN dataset [190] for genre identification is one of the most commonly used datasets for music classification. Unlike the Million Song Dataset, GTZAN provides audio samples, but is limited to around 1000 entries.

Smaller datasets for specific forms of world music are hosted by the Music Technology Group (MTG), at Universitat Pompeu Fabra.[19] Here, samples of styles such as Carnatic, Hindustani, and Turkish Makam Music are provided, along with various annotations. The datasets are generally between 100-500 samples in size. Audio sample libraries are also commonly available through a various licenses, and are extremely useful for a range of IMP tasks. Freesound[20] for example is also managed by MTG, and hosts a wide range of royalty free, Creative Commons files, whereas other libraries such as Jamendo, Spitfire and Loopmasters provide commercial alternatives.

Sound Synthesis and MIDI

Most modern audio workstations incorporate the use of software instruments, typically controlled using MIDI. Datasets that provide symbolic notation provide a means to test and train algorithms for intelligent tools which manipulate musical sequences.

The Lakh MIDI Dataset [191] is a collection of 176,581 MIDI files, with a subset of 45,129 files matched to entries in the Million Song Dataset and aligned to their corresponding MP3 files. The dataset is provided with metadata containing details such as artist, genre, year, and can be easily linked to previews and lyrics.

Similar datasets containing aligned audio and MIDI files tend to be popular in the field of MIR, as they are widely used for the alignment of audio to a symbolic representation of the score. These tend to be from the classical music genre, with one of the most popular being the Bach10 dataset [192]. Similarly, MIDI files have been generated using a subset of the MedleyDB and Bach10 datasets [193]. Here, an analysis/synthesis sinusoidal modeling framework is used for multiple-f_0 estimation. As a result, matched audio and MIDI files are produced. The dataset contains a number of different synthesis types, with over a hundred full tracks included in total. Finally, NES-MDB [194] is a dataset of 5278 MIDI files taken from the soundtracks of 397 games developed for the Nintendo Entertainment System. The data is extracted from the original files, which were represented in the Video Game Music (VGM) format, from which expressive timings and velocities were derived.

---

[18] https://labrosa.ee.columbia.edu/millionsong
[19] https://compmusic.upf.edu/datasets
[20] https://freesound.org

5.2 Data Collection Methods

5.2.1 In the DAW

DAW Session Analysis

Given the scarcity of processed stems and mixes with parameter settings, most studies analyze the mixing process in a lab setting, where a constrained set of commercially available or custom-built processors are operated by amateur or expert engineers [119, 123].

To address this limitation of mixing research, De Man et al. studied the practices and preferences of mix engineers by extracting and analyzing audio from DAW sessions, created specifically for the purpose of the experiment by trained mix engineers in realistic settings [74, 82, 178]. By allowing the creators to use the environment and professional-grade tools they are used to, the resulting mixes were maximally ecologically valid, while still enabling detailed analysis. Some restriction was placed on which plugins could be used, in coordination with the participants, so that the entire mix could be recreated after the fact by the researchers.

Compared to typical lab-based studies, this approach has the advantage of respecting the multidimensional nature of the mixing process. Signal processing tools may be used very differently by a novice user in a lab with no other processing available, from how a professional would use the same tool as part of a comprehensive mix process. Furthermore, different tools may be used to achieve similar goals, such as controlling a perceived level by applying gain, equalization or dynamic range compression.

The creation and assessment of realistic mixes is a laborious process, where some control is sacrificed for high ecological validity and preservation of the possible interplay of mix processes. This leads to a very wide scope, as all common mix processes may be included. Manual comment annotation of free-form text responses was also deemed necessary, as the vocabulary and salient perceptual dimensions of the mixing process are largely unknown. As a consequence, the approach is only moderately scalable, even when the data, tools and methods presented here can be repurposed. However, this type of exploratory studies may ultimately inspire more focused and controlled experiments, zooming in on a single aspect with proven relevance and known perceptual attributes.

Audio Plugins

To capture audio effect data, the aforementioned SAFE Project [184] embeds a data collection system into the music production workflow, by providing a suite of open source audio effects plugins. Users of the freely downloadable DAW plugins are able to submit descriptions of their audio transformations, while also gaining access to a continually growing collaborative dataset of musical descriptors. In order to provide more contextually representative timbral transformations, the dataset is partitioned using metadata, obtained within the application.

Each plugin UI consists of a parameter set, a visualization of the effect (e.g., a transfer function), and a free-text field, allowing for the user to input one or more text labels representing human-readable descriptions of the effect's processing (e.g., *warm*, *bright*, *fuzzy*). The descriptions are uploaded anonymously to a server along with a time-series matrix of audio features extracted both pre- and post-processing, a static parameter space vector, and a selection of metadata tags. To motivate the user base to provide this data,

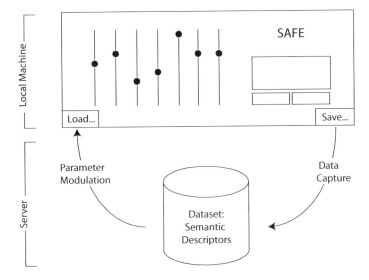

Figure 5.4 A schematic representation of the plugin architecture, providing users with load and save functionality, taken from [184].

parameters settings from a wider audience can also be loaded from the server, which allows access to a wide range of plugin presets (see Figure 5.4).

Four audio effect plugins have been implemented in VST, Audio Unit and LV2 formats, which comprise an amplitude distortion effect with tone control, an algorithmic reverb based on the figure-of-eight technique proposed by Dattorro [195], a dynamic range compressor with ballistics, and a parametric EQ with three peaking filters and two shelving filters. All visible parameters are included in the parameter space vector, and can be modulated via the text input field. The Graphical User Interfaces (GUIs) of the effects[21] are shown in Figure 5.5.

When users save a described transform, an $N \times M$ matrix of audio features is extracted before and after the transform has been applied, where N is the number of recorded frames and M is the number of audio features, which are captured using the LibXtract feature extraction library [196]. Furthermore, an optional metadata window is provided to store user and context information, such as the genre of the song, musical instrument of the track, and the user's age, location and production experience.

5.2.2 Via the Web

When gathering data either through analysis of recording sessions, or through the development of tools that extract information from a user's current project, there is a prerequisite that the user must have the relevant software, and often that the researcher or data collection team must be present. An alternative is to build software tools that can be used arbitrarily by subjects to capture and send data, irrespective of the system they are working on, or their geographical location. For this reason, it seems sensible to consider methods

---

[21] Plugins can be downloaded from www.semanticaudio.co.uk/projects/download

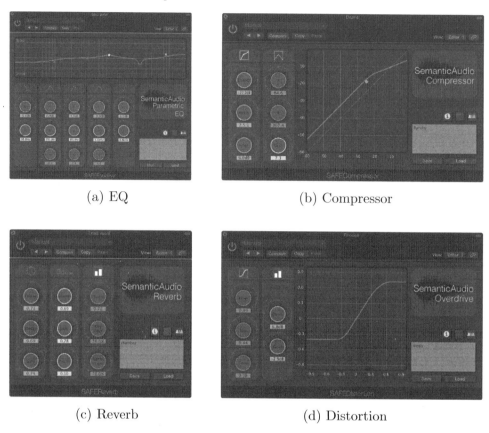

(a) EQ

(b) Compressor

(c) Reverb

(d) Distortion

Figure 5.5 Graphical user interfaces of the equalization, compression, reverberation, and distortion plugins

that utilize the web to not only transmit data, but also to allow users to record data using a browser.

Historically, web-based data collection methods have been frowned upon due to the potential for uncontrolled variability and biases caused by the environment, or subject group. There is however an often necessary trade-off when collecting large datasets, due constraints on time, space and cost. In the past, a number of methods have been deployed to maintain some degree of control over the subjects. For example, the MUSHRA listening test methodology [197–199, 224] has a failsafe for mitigating noisy data. The standard prescribes hidden reference and anchor samples, which should be rated most favorably and negatively, respectively. Similarly, Woods et al. [200] go as far as developing a listening test, which can be required pre-trial, to determine whether a subject is wearing headphones. This involves the subject listening to three sinusoidal test-tones, one of which is phase-inverted across the left and right channels. They are then asked to identify the quietest tone – a task that should be relatively easy when wearing headphones, but difficult over loudspeakers.

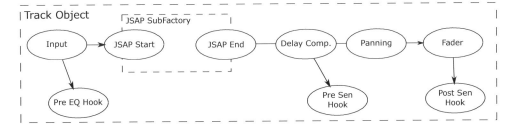

Figure 5.6 Structure of the track object and bus object in the audio engine

DAW Session Analysis

Due to the overhead of conducting controlled tests, DAW session data that can be collected remotely is extremely desirable for the IMP community. For this reason, a system developed by Jillings et al. [120] allows for a fully functional DAW to be deployed on the web by harnessing the features of the Web Audio API. By restructuring the standard host architecture, the DAW is specifically designed for the collection of semantics data, and the implementation of intelligent processing systems. The DAW is powered by an audio engine that supports sequencing, mixing, routing, and audio effects processing. In the system, all aspects of the audio production environment are classed as objects with a finite set of variables (based on [57]), and can be captured with their predefined relationships to other objects. Each object can be tracked back to the root *session* node and can be represented in a relational format, in which a Session owns a channel, which holds a region, and a processing chain, and so forth. The DAW is specifically designed for adaptive and cross-adaptive audio processing, via the JavaScript Audio Plugin (JSAP) framework [201], in which the routing of features and parameters at multiple points in the processing chain is enabled, as shown in Figure 5.6.

Audio Plugins

As the web is an emerging platform for audio processing, the standardization of a plugin framework such as VST or AudioUnit has yet to happen. The JSAP intelligent audio plugin framework [201] allows for the implementation of audio effects, with an emphasis on data-driven intelligent audio processing.

The core modules of the standard are the Base Plugin, in which the audio graph and connected interface are encapsulated, and the Plugin Factory, in which the host is connected to the processing modules currently being implemented. This structure enables cross-adaptive processing, semantic interaction and feature sharing. The role of the PluginFactory is to hold prototype objects for the creation of a processing chain as well as providing a single point of reference for the rest of the application.

The PluginFactory provides links to data stores, either for data collection or requesting. In doing so it enables plugins to directly communicate with connected web and semantic web databases. The factory also manages inter-plugin communications, enabling complex cross-adaptive processing by feature-driven modulation.

Feature extraction is embedded into the framework for real-time data acquisition, using the JS-Xtract [202] library. In JASP, designing auto-adaptive effects is possible by querying the previous plugin for features and using them internally, or placing a Web Audio API

AnalyserNode inside the plugin sub-graph. The PluginFactory can be fed global semantic information about the session such as tempo, sample rate and any user or personal information.

5.3 Data Representation

5.3.1 Formats

Multitracks

Raw uncompressed audio signals are encoded and stored digitally using Pulse Code Modulation, in which the signal is sampled at regular intervals, which are spaced equally in time, and quantized to a predetermined set of amplitude stages. PCM audio signals can be distributed in a number of formats, all of which represent the file's metadata in various ways. The RAW format for example includes no additional data relating to the file's parameters, including sample rate or bit depth. One of the most common formats for PCM audio data is Wave (.wav), an instance of the Microsoft's RIFF framework in which the header information is constructed in chunks. Each chunk begins with an identifier and a size block, allowing the file-reader to distinguish metadata such as sample rate, bit depth and number of channels through the format (fmt) chunk and from the audio data. One of the limitations to the wave format is that it cannot transmit multiple related multichannel tracks, all with individual format information and data. For this reason, most multitrack mixes are distributed using an archive containing multiple stems. This can present issue regarding alignment if recordings were started at various points in time, and often means it is necessary to distribute a DAW session file, or some form of file metadata.

The Stems format[22] from Native Instruments aims to get around this issue by providing an open multichannel format, which encodes four separate streams containing different musical elements in a modified MP4 file format. The file allows bass, drums, vocals and melody components to be isolated and remixed, and can be packaged in a single file. The format is primarily aimed at DJs, however it is already being utilized by a number of datasets, including the aforementioned MUSDB18 dataset, and is supported by the Traktor DAW.

Metadata

Other than sample rate, number of channels and bit depth, it is relatively uncommon to store metadata within the file itself. This is due to the lack of standardization of audio descriptors, and in the drive for more compact audio formats, audio features are edged out as they are not necessary for standard playback.

One of the most common frameworks for storing audio metadata is the HDF5 format,[23] which allows cross-platform extraction, with fast data I/O and heterogeneous data storage. The format is used by a number of audio datasets including the Million Song Dataset [189], specifically because it allows for the efficient storage of large variable-typed inputs such as differing-length audio feature vectors, longitude and latitude coordinates, and strings representing file names, artists and genres, among others. The file type is very common in academic research, and works using a hierarchical grouping structure, in which a group contains a subset of datasets. A dataset in this context is a multidimensional

---

[22] www.native-instruments.com/en/specials/stems
[23] www.hdfgroup.org/solutions/hdf5

array of data elements, each with a name tag, and a predefined list of properties with datatypes and space. The files are not stored in readable form, but can be read and written through a number of APIs in various programming languages such as C++, Python and MATLAB.

Another common format for metadata distribution is JavaScript Object Notation (JSON), which is distributed in human-readable format and can be accessed in a large number of languages, through numerous high-level APIs. The format is used by a number of datasets including the Lakh MIDI Dataset,[24] due to its compliance with web formats and extensibility. The format consists of key-value pairs, in a non-iterable unordered list, which is able to store variable-typed data in human-readable format. The format is popular with smaller datasets, particularly those that are used on the web. An example of a track metadata is given in the following code segment:

Listing 5.1 Example metadata stored in JSON format

```
{
    "song": {
        "title": "Heros",
        "artist": "David Bowie",
        "genre": "Alternative rock",
        "year": 1977
    },
    "signal": {
        "channels": 2,
        "samplerate": 44100
    },
    "features": {
        "centroid": [10341.75, 6581.03, 3724.45, 8251.84,
            10332.87],
        "rms": [0.24, 0.95, 0.42, 0.75, 0.26]
    },
    "timeStamps": [9, 32, 93, 114, 212]
}
```

Markup languages such as XML are also commonly used for storing metadata, based on the same principles of JSON. XML is used for distributing user data in the SAFE Dataset [184] as there are a number of high-level APIs for writing relational database stores. The language is similarly structured to that of JSON and adheres to the SGML standard.

The YAML (YAML Ain't No Markup Language) data-serialization language is also popular with audio datasets, including MedleyDB [169]. The format is human-readable and is commonly used for configuration files. YAML specifically breaks the SGML standard, and is intended to be an alternative to languages like XML with more simplistic notation. The format utilizes indentation to indicate nesting, and permits mappings such as hashes and dicts, sequences such as arrays and scalars such as strings and variable-typed numbers. The YAML equivalent of Listing 5.1 is given in Listing 5.2.

---

[24] Primarily used for matching MD5 checksums to original filenames

Listing 5.2 Example metadata stored in YAML format

```
song:
    title: Heros
    artist: David Bowie
    genre: Alternative rock
    year: 1977
signal:
    channels: 2
    samplerate: 44100
features:
    centroid:
    - 10341.75
    - 6581.03
    - 3724.45
    - 8251.84
    - 10332.87
    rms:
    - 0.24
    - 0.95
    - 0.42
    - 0.75
    - 0.26
timeStamps:
- 9
- 32
- 93
- 114
- 212
```

Databases

It is common practice for data to be stored in a relational database format, particularly when used as the model in a UI architecture (e.g,. model-view-controller, or model-view-presenter). Relational databases organize data into tables comprising records (rows) and attributes (columns), where entries are tuples. Most mainstream database formats utilize Structured Query Language (SQL). Some of the more common database management systems include MySQL,[25] PostgreSQL,[26] and SQLite[27] but these databases come in many forms and a detailed discussion is beyond the scope of the book.

These database architectures have been adapted in recent times to accommodate specific data properties. NoSQL databases for example do not rely on SQL and support more flexible query structures, typically to accommodate big data. This can be in the form of key-value datastores, document stores, or graph-based stores. These are typically designed for optimal performance and scalability due to the volume of some datasets, particularly in the fields of machine learning and data science.

[25] www.mysql.com
[26] www.postgresql.org
[27] www.sqlite.org

A further modification to the relational format is the triplestore system, which is intended to store triplets in a subject-predicate-object format, as opposed to tuples. This adheres to the Resource Description Framework (RDF) specification, which is designed for conceptual information modeling through the storage of relationships. Due to the comprehensive nature of the format, it is particularly useful for modeling ontological knowledge data, as discussed in Section 5.3.2. As opposed to SQL, databases of this kind use the SPARQL query language, which deals specifically with RDF data. Popular triple store models include AllegroGraph[28] and Stardog.[29]

5.3.2 Ontologies and Knowledge Representation

An ontology is a semantic data model, in which domain knowledge is formalized, to provide a rigorous underlying set of relationships between objects [203]. Ontologies are at the core of a wide range of technologies behind the semantic web, as they aim to quantify knowledge and represent it in an extensible, homogeneous format. Ontology languages such as OWL (Web Ontology Language) [204] and frameworks like RDF enable ontological concepts to be encoded, which in turn are stored in SPARQL-based data stores. These form the infrastructure for a network of interlinked uniform resource identifiers (URIs), in which knowledge can be formalized, distributed and reasoned upon by computational agents. In audio engineering a number of key ontologies are relevant to the representation of production data, some of which are being used extensively for the storage, representation and processing of digital media.

The Music Ontology
provides an architecture for the representation of musical data [205]. The Music Ontology extends the functionality of a number of other ontologies, namely the Timeline Ontology,[30] which allows temporal data to be stored using RDF, and the Event Ontology,[31] which provides representations of discrete occurrences that occupy arbitrary time/space regions. In order to specifically represent artistic works, the Functional Requirements for Bibliographic Records ontology (FRBR) [206] is used, whereby a work may refer to a composition, and a manifestation may refer to the medium on which it was released.

The Studio Ontology
provides a framework for capturing sound production data [207]. Here, concepts from the production workflow and the studio environment such as equipment, metadata and signal processing are embedded in RDF format. The Studio Ontology extends a number of other frameworks including the Connectivity Ontology, and the Music Ontology as shown in Figure 5.7.

The Studio Ontology provides a representation of audio effects, and their application in sound production, such as: (1) *foundational components* which allow for the characterization of technological artifacts, (2) *complex device descriptions* which represent concepts such as signal processing tools and their relationships to other components in session, (3) *core components* such as devices, equipment and staff, and (4) *domain specific extensions*.

---

[28] https://allegrograph.com
[29] www.stardog.com/platform
[30] http://motools.sourceforge.net/timeline/timeline.html
[31] http://motools.sourceforge.net/event/event.html

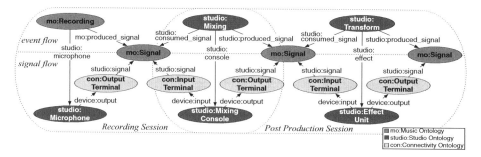

Figure 5.7 Outline of The Studio Ontology, presented in [207]

The Audio Effects Ontology

provides detail about the parameterization and classification of audio processing devices, which are omitted from broader ontologies such as the Studio Ontology [208, 209]. It defines audio effect implementations and transformations and introduces different effect classification systems, such as those presented in Chapter 2. It also provides concepts that allow the specification of products and product families, used to represent audio effects in a commercial scenario. The ontology consists of the following modules: (1) the *Main Ontology*, which includes general effect specification such as details on parameter settings, (2) a *Parameter Ontology*, which encapsulates parameter descriptions and types, and (3) an *Effect Classification Taxonomy*.

The Audio Feature Ontology

is an extension of the event and timeline ontologies with a focus on providing more enriched metadata relating to the extraction of features from audio data [210]. The ontology addresses the requirement to capture data which denotes not only the feature vectors themselves, but also information regarding the methods in which they were extracted. This accounts for issues such as discrepancies in feature calculations due to feature extraction libraries using different algorithms, or from differing time stamps, sample rates, etc. The ontology is structured in a way that represents a range of feature abstractions, from the conceptualization of the feature (i.e., its design and motivation), through to the implementation of the feature extraction, on a specific computational device. In the ontology, objects are represented hierarchically, where feature concepts such as the Spectral Centroid, or Kurtosis are of an audio signal are subclasses of Audio Features, which in turn are events, on a timeline.

The Semantic Audio Feature Ontology

was designed to describe the semantics relating to the application of an audio effect and the audio feature of the signals involved [211]. It is based on the Audio Effects Ontology, for use with the data collected during the SAFE project [184], introducing additional concepts for the human annotation and description of audio processing techniques. This additional data can be separated into three groups:

- Metadata describing the application of an audio effect.
- Audio feature data for the input and output signals.
- Provenance of the data.

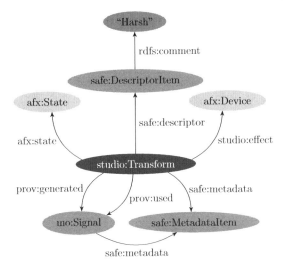

Figure 5.8 The structure used to describe the application of an audio effect

Each entry in the SAFE dataset is described using the *studio:Transform* concept, applying some semantic transformation on a set of input signals to produce a set of output signals. The structure of this data is shown in Figure 5.8.

The SAFE Ontology makes extensive use of the Provenance Ontology in order to identify the origins of the data.

5.3.3 *Licenses*

Particularly with datasets containing audio files, licensing is extremely important, as it dictates the possible applications in which the data can be used. For academic purposes, the use of audio data tends to be a gray area, particularly if that data then goes on to form the basis of any commercial projects, either through algorithm training, or through the derivation of performance rules. Creative Commons (CC) licensing enables content to be distributed under a set of standardized flexible licenses, with a varying range of stipulations, from simple creator attribution, to non-commercial use. The CC license itself has three layers: a legal code layer, a human-readable commons deed, and a machine-readable layer.

There are six main Creative Commons licenses. In order of least to most restrictive, they are:

- **Attribution (CC BY)**: the only stipulation is that the work is credited to the original author.
- **Attribution – Share Alike (CC BY-SA)**: in addition all following works must be distributed under the same CC license as the original work.
- **Attribution – No Derivative (CC BY-ND)**: in addition, no permission is given to change the original content in any way. This includes remixing, editing or repurposing.

- **Attribution – Non-commercial (CC BY-NC)**: nobody else is permitted to make money from the works other than the original author.

- **Attribution – Non-commercial – Share Alike (CC BY-NC-SA)**: users are restricted to only use the content for non-commercial applications, and future works must be distributed under the same license

- **Attribution – Non-commercial – No Derivatives (CC BY-NC-ND)**: users are restricted to non-commercial use, and no derivatives can be made of the original content.

Based on the CC notion of reusability, the Audio Commons project was recently initiated with music sharing in mind [212]. The project focuses on the fact that the web has an extensive range of Creative Commons audio content, and an increasingly large amount of public domain audio, due to expired licenses, but that there are no mechanisms to efficiently search disparate repositories and find relevant content. Furthermore, there is still widespread confusion over the mechanisms that should be used for reuse, and relicensing. The project incorporates huge sample and song repositories such as Freesound and Jamendo and boasts a wide range of commercial and academic partners.

The Audio Commons Ecosystem proposes a wide range of tools for content creators, providers and users, through a range of APIs and ontologies [213]. This encourages widespread use of content under flexible licenses, with formalized processes, to annotate, search and share Creative Commons audio data.

6

Perceptual Evaluation in Music Production

"There is not a question that cannot be addressed, that can't be answered, or at least discussed, with critical listening. Critical listening tells you everything you need to know."

George Massenburg in "Parametric Music,"
Pro Sound Web, 2012.

At the end of the day, a mix, or mixing system, is only as good as the subjective impression it leaves on its audience. In this regard, music mixing is different from most other topics within engineering, where system requirements are usually well-defined, and can be expressed in simple metrics such as speed, strength or power. Therefore, a good understanding of the perceptual effects of the various mix processes is paramount if a successful mixing system is to be designed, and a mix can only be validated by subjective evaluation. However, many aspects of mix engineering remain poorly understood with regard to their impact on perception, so more research is needed in this area.

Central to any study on perception is a listening test, in which multiple subjects indicate their overall preference or a more specific perceptual attribute with regard to one, two or even a dozen stimuli at a time. In some cases, a 'method of adjustment' is used where a subject's manipulation of some parameter of the audio are recorded. An effective methodology helps produce accurate results, with high discrimination, and minimal time and effort. In what follows the measures necessary to accomplish this are investigated. Whereas great resources about subjective evaluation of audio already exist [214], we only consider the assessment of music production practices and processes. This presents strategies for the perception-centered design and validation of intelligent music production systems, and offers an overview of examples of IMP evaluation.

6.1 Basic Principles

Certain principles are essential to any type of listening test, and are supported by all known software tools [214]. They relate to minimizing the uncontrolled factors that may cause ambiguity in the test results [197]. While some are a challenge to accommodate in an analog setting, software-based listening tests fulfill these requirements with relative ease.

First, any information that should not distract the subject from the subjective task at hand should be concealed, including any metadata regarding the stimulus [215]. In other words, the participant should be 'blind.' Furthermore, the experimenter should not have any effect on the subject's judgment, for instance by giving subconscious cues through facial expressions or body language. This is commonly referred to as the double blind principle, and is easily achieved in the case of a software-based test when the experimenter is outside of the subject's field of view or even the room.

A subject can be similarly biased by the presence of other subjects who are taking the test at the same time. For this reason, it is advised that the test is conducted with one person in the room at a time, so as not to influence each other's response.

Another potential bias is mitigated by randomizing the order in which stimuli are presented, as well as the order of the pages within a test, and the order of entire test sessions if there are several. This is necessary to avoid uneven amounts of attention to the (sets of) stimuli, and average out any influence of the evaluation sequence, such as subconsciously taking the first auditioned sample as a reference for what follows [197, 216]. In case a limited number of subjects take part, a pseudo-random test design can ensure an even spread over the different 'blocks,' e.g., in the case of two sets of stimuli, 50% of the subjects would assess the first set last.

6.2 Subject Selection and Surveys

In the following, a distinction is made between *skill*, i.e., experience in audio or music, and *training*, i.e., preparing for a specific test, for instance by including a training stage from which results are not used for analysis. Therefore, a subject can be skilled on account of being an audio professional, but untrained due to lack of a training stage preceding the listening test.

Results from more skilled subjects are likely to have higher discrimination [216–218], higher agreement [139], and to be more reliable [197]. The same can be said for trained subjects. However, training is costly both in terms of time and materials, as responses from a training phase are not usually regarded as valid results. Instead, the order of the pages can be logged, so that it is possible to omit the results of the first part of the test.

To avoid misunderstandings, the experimenter should give detailed instructions, ideally in different formats. For instance, it is customary that written instructions are available, but that these are also communicated verbally [216]. In addition, subjects can be reminded of the tasks through on-screen text, before the test and between different pages. The risk of worthless test sessions is further reduced by implementing checks that warn subjects when the instructions are not followed. For instance, alerts can be shown when a stimulus was not played, rated, or commented on. In order to still be able to use repeated failure to execute certain tasks as a basis to exclude participants, all of these events should be logged. Finally, the interface should make the evaluation as straightforward and logical as possible, so that the task would be clear even without further guidance.

Exclusion of a certain subject's results can be deemed necessary based on self-reported hearing problems, indications of misunderstanding the assignment, strong deviation from other subjects, failure to repeat own responses, or incomplete data. When one or more stimuli are not evaluated by a subject, the ratings of the remaining stimuli are not necessarily comparable to the other ratings and may need to be discarded. Of course, every such exclusion should be reported and motivated, to avoid perceived cherry-picking [216].

Apart from hearing impairments, a post-test survey may also be used to keep track of subject age, gender, nationality, native language, experience with audio or music, and general comments about the experiment. This data may be relevant to report on, but it may also help discover issues with the methodology or lead to interesting findings.

6.3 Stimulus Presentation

6.3.1 Pairwise or Multiple Stimuli

When selecting the appropriate interface, a first important distinction is between single stimulus interfaces, where one stimulus is evaluated at a time; pairwise interfaces, where the subject assesses how two stimuli compare to each other; and multi-stimulus interfaces, where more than two stimuli are presented 'at the same time,' for the subject to compare in any order.

In case at least two differently processed versions of the same material are presented simultaneously, subjects are likely to focus on the contrasting sonic properties rather than the inherent properties of the source [219]. As this is the desired effect for the purposes of this field, the single stimulus approach is ruled out: not the properties of the source material, but the differences in processing should be evaluated.

Many researchers have previously considered the performance of pairwise versus multi-stimulus tests, and judged that the latter are preferable as long as the number of conditions to be compared is not too large – preferably lower than 12 [220] or 15 [197] – as they are more reliable and discriminating than both pairwise and single stimulus tests [217, 221]. In the case of attribute elicitation, multi-stimulus presentations enlarge the potential pool of descriptors, without the high number of combinations required in the case of pairwise comparison [222]. The number of stimuli should be as high as possible without making the task too tedious, as this elicits a richer response [223].

6.3.2 To MUSHRA or not to MUSHRA

The Multi Stimulus test with Hidden Reference and Anchor (MUSHRA) [197] is a well-established type of test, originally designed for the assessment of audio codecs, i.e., the evaluation of (audible) distortions and artifacts in a compromised signal. Some of the defining properties of the associated interface, set forth by Recommendation ITU-R BS.1534-1 (also referred to as the MUSHRA standard), are:

- multiple stimuli, at least 4 and up to 15, are presented simultaneously;
- a separate slider per stimulus, with a continuous quality scale marked with adjectives 'Bad,' 'Poor,' 'Fair,' 'Good' and 'Excellent';
- attributes to be rated can be one or more, but should include 'basic audio quality';

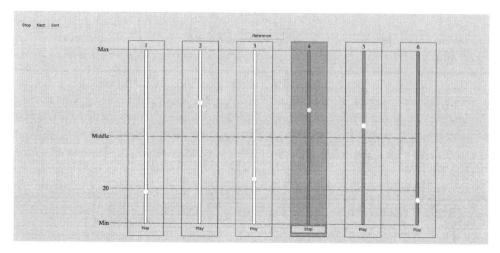

Figure 6.1 Example of a MUSHRA interface, implemented with the free Web Audio Evaluation Tool [224]

- a reference stimulus is provided;
- the reference is also included in the stimuli to be rated, as a 'hidden reference'; and
- among the stimuli to be rated are one or more low-quality 'anchors.'

Despite being developed with codec evaluation in mind, MUSHRA-style interfaces have been used for many other purposes, including evaluation of mixes [46,49,63,225]. They have the advantage of being well-known and well-defined, so that they need little introduction in the context of research papers, and results from different studies can be compared. However, some question its suitability for non-codec applications, and deviate from the rigorous MUSHRA specification to better address the needs of the research at hand. In this section the argument is made to employ a different test interface for the subjective evaluation in music production research.

First and foremost, a 'reference' is not always defined [136]. Even commercial mixes by renowned mix engineers prove not to be appropriate reference stimuli, as they are not necessarily considered superior to mixes by less experienced engineers [178]. The MUSHRA standard itself specifies that it is not suitable for situations where stimuli might exceed the reference in terms of subjective assessment. It is aimed at rating the attribute 'quality' by establishing the detectability and magnitude of impairments of a system relative to a reference, and not to measure the listeners' preference [226].

The inclusion of a purposely very low-quality 'anchor' sample tends to compress the ratings of the other stimuli, which are pushed to the higher end of the scale, as the large differences the anchor has with other stimuli distract from the meaningful differences among the test samples. An anchor serves to assess the ability of test participants to correctly rate the samples, to enforce a certain usage of the scale (i.e., to 'anchor' the scale at one or more positions), and to indicate the dimensions of the defects to identify. As the task at hand is a subjective one, and the salient objective correlates of the subject's perception are not known, this is not applicable here. A final drawback of including anchors is the increased number of stimuli to assess, raising the complexity and duration of the task.

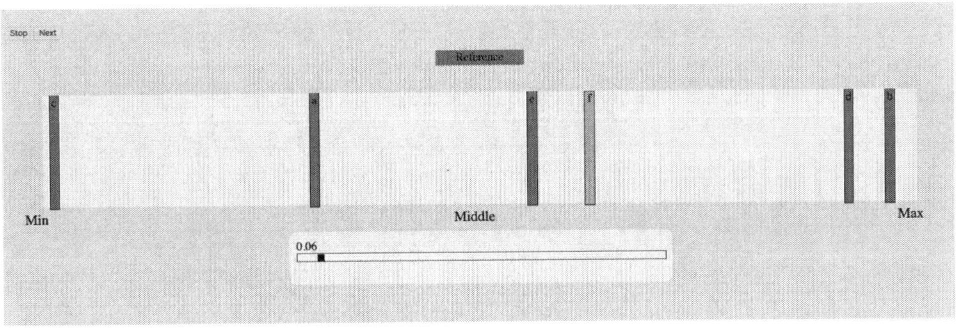

Figure 6.2 A single-axis, multi-stimulus interface, based on [228], implemented with the free Web Audio Evaluation Tool [224].

In the absence of anchors and references, and as listeners may or may not recalibrate their ratings for each page, resulting ratings cannot be compared across pages, though the average rating likely reflects their overall liking of all processed versions and the source material itself [227].

MUSHRA tests prescribe a separate slider per stimulus, to rate the quality of each individual sample relative to a reference. In the present context, however, no reference is defined, and the perception between the different stimuli is of interest. A single rating axis with multiple markers, each of which represents a different stimulus, encourages consideration of the relative placement of the stimuli with respect to the attribute to be rated [228]. A 'drag and drop' type interface, such as the one depicted in Figure 6.2, is more accessible than the classic MUSHRA-style interface, particularly to technically experienced subjects [198]. It also offers the possibility of an instantaneous visualization of the ranking, helping the assessor to check their rating easily, and making the method more intuitive. As stated in the MUSHRA specification itself, albeit as an argument for multi-stimulus as opposed to pairwise comparison, the perceived differences between the stimuli may be lost when each stimulus is compared with the reference only [197]. This means that while a true 'multi-stimulus' comparison of test samples, where each stimulus is compared with every other stimulus, is technically possible with separate sliders (even without a reference), it is probable that a subject may then not carefully compare each two similar sounding stimuli.

6.3.3 *Free Switching and Time-aligned Stimuli*

Rather than playing all stimuli in a randomized but fixed sequence, allowing subjects to switch freely between them enhances the ability to perceive more delicate differences [229]. While this is fairly ubiquitous in digital listening test interfaces, some older experiments did not accommodate this.

The comparison of differently processed musical signals is further facilitated by synchronized playback of time-aligned audio samples, and rapid, immediate switching between them. This leads to seamless transitions where the relevant sonic characteristics change instantly while the source signal seemingly continues to play, directing the attention to the differences in processing rather than the intrinsic properties of the 'song.' It also avoids

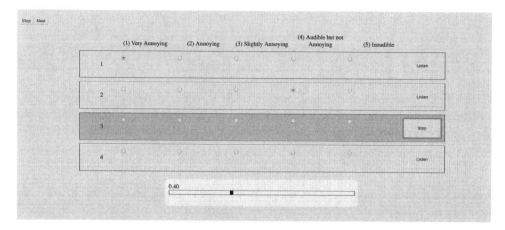

Figure 6.3 A discrete, multi-stimulus interface, implemented with the free Web Audio Evaluation Tool [224]

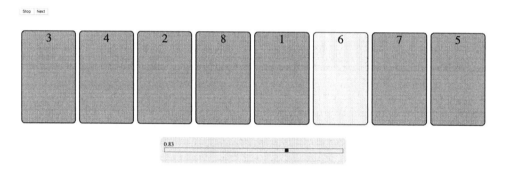

Figure 6.4 A simple ranking interface, implemented with the free Web Audio Evaluation Tool [224]

excessive focus on the first few seconds of long stimuli, and makes toggling between them more pleasant.

6.4 Response Format

6.4.1 Discrete and Continuous Rating and Ranking

In the case of a discrete scale (Figure 6.3), it would for instance be possible for a subject to rate each sample in a test as 'Good,' providing very little information and therefore requiring a high number of participants to obtain results with high discrimination. Conversely, subjects may feel compelled to rate two tracks differently (e.g., 'Good' and 'Excellent') even when they hear very little difference.

A plain ranking interface (Figure 6.4) is not advisable either, as it would prevent learning which stimuli a subject perceives as almost equal, and which are considerably different.

Thus, a continuous scale is most appropriate for the application at hand, as it allows for rating of very small differences. Tick marks should be omitted, to avoid a buildup of ratings at these marks [230].

6.4.2 Rating Scale Names

Scale names used in listening tests often appear to have been defined by the experimenter, rather than derived from detailed elicitation experiments, and are therefore not necessarily meaningful or statistically independent of each other [231]. Scales associated with specific, fixed attributes further suffer from several biases, from a 'dumping bias' when 'missing' attribute scales impact the available scales [232], to a 'halo bias' when the simultaneous presentation of scales causes ratings to correlate [214]. Furthermore, the terms used may be understood differently by different people, particularly non-experts [229]. No established set of attributes exists for the evaluation of music production practices, whereas literature on topics like spatial sound includes many studies on the development of an associated lexicon [231, 233, 234]. As such, instead of imposed detailed scales, the use of one general, hedonic scale is recommended. Exceptions exist, such as when the study is preceded by a description elicitation phase to determine relevant attributes to rate.

Evaluation of audio involves a combination of hedonic and sensory judgments. Preference is an example of a hedonic judgment, while (basic audio) quality – "the physical nature of an entity with regards to its ability to fulfill predetermined and fixed requirements" [235] – is primarily a sensory judgment [236, 237]. Indeed, preference and perceived quality are not always concurrent [136, 238, 239]: a musical sample of lower perceived quality, e.g., having digital glitches or a 'lo-fi' sound, may still be preferred to other samples which are perceived as 'clean' but don't have the same positive emotional impact. Especially when no reference is given, subjects sometimes prefer a 'distorted' version of a sound [240]. Unless the degree of perceived artifacts is being studied, 'preference' is therefore the appropriate scale label.

6.4.3 Comments

A single, hedonic rating can reveal which stimuli are preferred over others, and therefore which parameter settings are more desirable, or which can be excluded from analysis. However, it does not convey any detailed information about what aspects of a mix are (dis)liked. Furthermore, subjects tend to be frustrated when they do not have the ability to express their thoughts on a particular attribute.

For this reason, free-form text response in the form of comment boxes should be accommodated. The results of this 'free-choice profiling' also allow learning how subjects used and misused the interface, whereas isolated ratings do not provide any information about the difficulty, focus, or thought process associated with the evaluation task. A final, practical reason for allowing subjects to write comments is that taking notes on shortcomings or strengths of the different stimuli helps keep track of which fragment is which, facilitating the complex task at hand.

Prior to the experiment it is unknown which instruments, processors or sonic attributes draw the most attention, and whether the salient perceptual differences between differently processed musical signals can be expressed using descriptive terms (e.g., 'drums are uneven') or if more detailed descriptions are typical (e.g., 'snare drum is too loud'). For this reason, a maximally free response format is preferable. Subsequent, more focused studies aimed at constructing a vocabulary pertaining to specific processors or instruments will be useful for the successful development of high-level interfaces.

A separate comment box per stimulus yields more and longer comments than when only one general text box is available [228]. To minimize the risk of semantic constraints on subjects [241] and elicit the richest possible vocabulary [219], all subjects should be allowed to use their native tongue. This necessitates tedious, high-quality translation of the attributes [242], ensured in some cases by several bilingual experts on the topic [234]. However, it is understood that experienced sound engineers are most comfortable using terminology and audio descriptors in the language they work in, and that English terms are pervasive in audio engineering settings across many countries, regardless of their native language [139].

6.4.4 Method of Adjustment Tests

In some cases, an experiment is not a passive listening test – where the samples cannot be modified by the listener – but an active parameter setting exercise. In this case, other considerations have to be kept in mind. For instance, 'endless knobs' without markings or visual feedback and randomized starting positions mitigate the risk that subjects internalize a certain number of 'turns' to consistently set the same balance, instead of listening for the ideal setting [131]. For the same reason, fader-style controls should have randomized lower and upper bounds and initial position, or avoided altogether.

6.5 Set-up

6.5.1 Headphones or Speakers

While headphones represent an increasingly important portion of music consumption, they are usually regarded as a secondary monitoring system for mixing, when high-quality speakers in an acoustically superior space are available. The use of headphones may lead to a sensory discrepancy between vision and hearing, and differences in preferred mix attributes compared to speaker listening [134]. Since each channel is only heard by one ear without cross-talk to the other ear, 'phantom' center images are harder to localize [243, 244]. With the exception of binaural audio, which is not considered here, most sources in stereo music are generally positioned 'inside the head' when listening to headphones, which may not be a desired effect [245]. In addition, a full range loudspeaker is not only heard but also felt, which is not possible when listening to headphones [246]. With regard to accuracy, listening over speakers is similar [247] or better [244] than listening over headphones. When no sufficiently high-quality room is available, though, headphones are preferable as they are impervious to the acoustic properties of the space [244].

6.5.2 Listening Environment

An important prerequisite for critical listening is a quiet, high-quality, acoustically neutral listening environment [117]. This need increases as the differences between stimuli become more subtle. Such a room will have a flat frequency response (±3 dB between 20 Hz and 18 kHz) [131], a controlled reverberation time across all frequencies (within a certain variation of the optimum value), and a low noise floor (less than NR 10). To answer the difficult question "What is enough?", one can refer to standards like ITU-R BS.1116 [216] or refer to related works. As with everything, the most essential requirement is to extensively report on the environment and its relevant measures so that they may be taken into consideration when interpreting results.

6.5.3 Listening Level

When loudness should not have an influence on the rated attributes, the playback level of each stimulus should be adjusted to be equally loud [214, 248, 249]. To this end, the integrated ITU-R BS.1770 loudness [250] needs to be constant across stimuli, and ideally across pages.

Subjects can be instructed to first set the listening level as they wish, since their judgments are most relevant when listening at a comfortable and familiar level [251], and since many perceptual features vary with level, e.g., the perceived reverberation amount [252,253]. Some studies allow only a ±4 dB deviation from a reference level [216], while others set a fixed level for all subjects [254]. Even when the level setting can be manipulated by the user, these changes can be logged as an extra experimental variable.

6.5.4 Visual Distractions

A key principle in the design of auditory perception experiments is to keep visual distractions to a minimum. In the context of digital interfaces, this means only having essential elements on the screen, to minimally distract from the task at hand [216, 255], and to avoid the need for scrolling, improving the subjects' accuracy and reaction times [256]. Apart from the necessary rating scale, comment boxes and navigation buttons, a tradeoff needs to be made between the value added and the attention claimed by interface elements like progress indicators, a scrubber bar and additional explanatory text.

In some cases, even a computer screen is deemed too distracting, and the graphical user interface is shown on a small mobile device [131]. Visual distractions in the room need to be reduced as well. This can be accomplished in part by dimming the lights, and covering any windows [255].

6.5.5 Online Tests

In many situations, listening tests are run on one or more computers in dedicated listening rooms, sometimes in different cities or countries. As these computers may have different operating systems and versions of the necessary software, developing an interface that works on all machines can be a challenge. Furthermore, as new versions of such software may not support the tool, it is best to reduce dependencies to a minimum. When the test is run locally, a problem with the machine itself can lead to loss of all results thus far – including tests from previous subjects if these were not backed up. Result collection from several computers, especially when they are remote, is tedious and can easily lead to lost or misplaced data. Similarly, installation, configuration and troubleshooting can be a hurdle for participants or a proxy standing in for the experimenter.

All these potential obstacles are mitigated in the case of a browser-based listening test: system requirements are essentially reduced to the availability of certain browsers, installation and configuration of software is not needed, and results could be sent over the web and stored elsewhere. On the server side, deployment requirements only consist of a basic web server. As recruiting participants can be very time-consuming, and as some studies necessitate a large or diverse number of participants, web-based tests can enable participants in multiple locations to perform the test simultaneously [199].

Finally, any browser-based listening test can be integrated within other sites or enhanced with various kinds of web technologies. This can include YouTube videos as instructions, HTML index pages tracking progression through a series of tests, automatic scheduling of a next test through Doodle, awarding an Amazon voucher, or access to an event invitation on Eventbrite.

Naturally, remote deployment of listening tests inevitably leads to a certain loss of control, as the experimenter is not familiar with the subjects and possibly the listening environment, and not present to notice irregularities or clarify misunderstandings. Some safeguards are possible, like assertions regarding proper interface use and extensive diagnostics, but the difference in control cannot be avoided entirely. Note, however, that in some cases the ecological validity of the familiar listening test environment and the high degree of voluntariness may be an advantage [257]. While studies have failed to show a difference in reliability between controlled and online listening tests [247, 258], these were not concerned with the assessment of music production practices.

6.6 Perceptual Evaluation of Mixing Systems

Before they are ready for practical use, intelligent software tools need to be evaluated by both amateurs and professional sound engineers in order to assess their effectiveness and to compare different approaches. In contrast to separation of sources in multitrack content, there has been little published work on subjective evaluation of the intelligent tools for mixing multitrack audio. Where possible, prototypes should also be tested with engineers from the live sound and post-production communities to judge the user experience and compare performance and parameter settings with manual operation. This research would both identify preferred sound engineering approaches and allow automatic mixing criteria derived from best practices to be replaced with more rigorous criteria based on psychoacoustic studies.

One of the chief distinguishing characteristics between the early work on intelligent mixing systems and later work is that very few of the early systems had any form of subjective evaluation, whereas now this is standard practice (see Table 6.1). Some systems are evaluated objectively, which means a measurement is made showing the system achieves a quantifiable objective, when not self-evident. This stops short of demonstrating an aesthetic improvement or a lack of unwanted side-effects, both of which can be addressed by subjective evaluation.

Table 6.1 Overview of systems that automate music production processes

	Objective evaluation	Subjective evaluation	No evaluation
Single track	[37, 42, 51, 52, 59, 65]	[51, 53, 56, 66, 69]	[22, 60, 70]
Multitrack	[3, 24, 36, 38–41, 44, 45, 47, 50, 57, 58, 61, 259]	[43, 46, 48, 49, 54, 55, 58, 61–64]	[4, 30, 67, 68, 71]

Mansbridge et al. [46] compared their proposed autonomous faders technique with a manual mix, an earlier implementation, a simple sum of sources and a semi-autonomous version. In [49] they compared an autonomous panning technique with a monaural mix and panning configurations set manually by three different engineers. Both showed that fully autonomous mixing systems can compete with manual mixes.

Similar listening tests for the multitrack dynamic range compression system by Maddams et al. were inconclusive, however, since the range of responses were too large for statistically significant differences between means, and since no dynamic range compression was often preferred, even over the settings made by a professional sound engineer [48]. But a more rigorous listening test was performed by Ma et al., where it was shown that compression applied by an amateur was on a par with no compression at all, and an advanced implementation of intelligent multitrack dynamic range compression was on a par with the settings chosen by a professional [62].

In [64], Wichern et al. first examined human mixes from a multitrack dataset to determine instrument-dependent target loudness templates. Three automatic level balancing approaches were then compared to human mixes. Results of a listening test showed that subjects preferred the automatic mixes created from the simple energy-based model, indicating that the complex psychoacoustic model may not be necessary in an automated level setting application.

De Man and Reiss [54] presented a fully knowledge-engineered, instrument-aware system on a par with professional mix engineers, and outperforming a combination of implementations of earlier, instrument-agnostics systems [46, 49, 61, 62].

Pestana and Reiss compared a multitrack time-frequency bin panning effect to a traditionally panned mix and a monophonic 'anchor' in [58], in terms of 'clarity,' 'production value' and 'excitement.' Relative to the professionally mixed version, the effect scored similar on 'clarity,' lower on 'production value,' and highest on 'excitement.'

Scott and Kim applied a single set of instrument-dependent panning and gain values was applied to ten mixes, and subjectively compared to a 'unity gain' version [55]. This basic knowledge engineered approach was favored six out of ten times.

Giannoulis et al. used a method of adjustment test to compare the settings of amateur and professional mix engineers to those returned by the proposed algorithm [53].

To evaluate the automatic mastering EQ based on fundamental frequency tracking in [51], Mimilakis et al. asked subjects whether they preferred content processed by this effect over the original source content. This was the case close to 80% of the time.

They also evaluated an automatic dynamic range compression system for mastering, based on deep neural networks, alongside a professionally mastered mix as well as third party genre-dependent mastering system in [66].

In [69], Chourdakis and Reiss trained several machine-learning models to apply reverberation, and consequently rated with regard to how closely they approximate a reference setting taken from the same training data. Due to the nature of the task, a MUSHRA-interfaced was used.

One of the most exciting and interesting developments has been perceptual evaluation of complete automatic mixing systems. In [63], Matz et al. compared various implementations of an automatic mixing system, where different combinations of autonomous multitrack audio effects were applied, so that one could see the relative importance of each individual tool. Though no comparison was made with manual mixes, it was clear that the application of these combined tools results in a dramatic improvement over the original recording.

Part III
How Do We Perform Intelligent Music Production?

7

IMP Systems

"I've had so many shows I've played ruined by really bad sound mixers and seen so many shows that were ruined by a bad sound mix, that I welcome the idea."

Reaction on pro audio forum Gearslutz, to "Aural perfection without the sound engineer," *New Scientist*, January 2010.

Intelligent Music Production is at present a growing and exciting field of research and many commercial devices based on its principles have started to emerge. The tools described in this chapter mainly address the technical constraints and challenges of music production, and are meant to automate various existing technologies that ordinarily would be performed manually. This then allows the sound engineer to concentrate on more creative aspects of the mix.

The state of the art in *automatic mixing* was described by Reiss [80] in 2011, but since then the field has grown rapidly. Table 7.1 provides an overview of intelligent mixing systems since those described in [80]. These technologies are classified in terms of their overall goal, whether they are multitrack or single track, whether or not they are intended for real-time use, and how their rules are found.

7.1 Level Balance

7.1.1 Balance Assumptions

Except for vague guidelines ("every instrument should be audible," "lead instruments should be roughly x dB louder"), there is very little information available on exact level or loudness

Table 7.1 Classification of intelligent audio production tools

Single or multitrack	Audio effect	Reference	Real-time	Rules
Single track	Equalization	[52]	Yes	Mix analysis
		[186, 260–264]	No	Machine learning
		[51]	Yes	Best practices
	Compression	[53, 59, 60]	Yes	Psychoacoustics; Best practices
		[66]	No	Machine learning
	Reverberation	[65, 69, 187, 265, 266]	No	Machine learning
	Distortion	[56]		Best practices
Multitrack	Faders and gains	[38, 39, 46]	Yes	Best practices
		[41, 47, 50, 57, 64]	No	Best practices
		[68]	No	Machine learning
		[44]	No	Machine learning
	Equalization	[40, 61]	Yes	Best practices
	Compression	[48, 62]	Yes	Psychoacoustics; Best practices
	Stereo panning	[36, 43, 49, 58]	Yes	Best practices
	Reverberation	[70]	No	Best practices; Machine learning
	Full mix	[54, 55, 71, 155, 162]	No	Best practices; Machine learning

values from practical sound engineering literature. A possible reason for this is the arbitrary relationship between the fader position and the resulting loudness of a source, as the latter depends on the initial loudness and the subsequent processing [259]. It can not be characterized by mere parameter settings and engineers are therefore typically unable to quantify their tendencies through self-reflection. Whereas a source's stereo position is solely determined by a pan pot, and its spectrum is rather predictably modified by an equalizer, a fader value is meaningless without information on the source it processes. RMS level channel meters give a skewed view as they overestimate the loudness of low frequencies, and more sophisticated loudness meters are not common on channel strips in hardware or software. Other factors may contribute to the absence of best practices, such as a dependence on genre or personal taste.

Some automatic mixing strategies have been based around the assumption that each track should be equally loud [39, 46]. This assumption introduces some conceptual issues: does one make every instrument equally loud, every microphone, or every group of similar instruments? For instance, should a drum set recorded with eight microphones be louder than one recorded with two microphones, and should a band with three guitars have the guitar loudness three times as high as for a band with a single guitar?

The assumption of equal loudness has been disproved by several studies [12, 64, 74, 120, 123, 267, 268]. The loudness measure typically used in this context is the ITU-R BS.1770 loudness [250]. Instead, analysis and perceptual evaluation of mixes has suggested that the vocal should be approximately as loud as everything else, i.e., the vocal should be 3 LU below the total mix loudness [12, 74, 123, 267].

Jillings and Stables [120, 268] investigated this same question with an online digital audio workstation that captured mix session information. In 71 sessions where subjects were asked to produce a balanced mix, vocals were again given dominance. However, lead vocals were, on average, set equally loud as the bass guitar, and 11 LU down from the total mix loudness.

This was far less than what the other studies found. Wichern et al. [64] also found that lead vocals were the loudest source, but this time approximately 7 LU lower than the total mix. The study also used other measures of loudness and partial loudness from Moore and Glasberg [269, 270], and found similar behavior, albeit on different scales.

In a series of papers [131, 132, 244, 271, 272], Richard King and co-authors investigated the preference for level balance. In [131], mixing engineers were presented with a stereo backing track and a solo instrument or voice. Their task was to adjust the relative levels when mixing the two, performed over a number of trials. Results showed that variance over trials was relatively low, and even lower for the more experienced subjects. They also showed some genre-dependent differences, which were investigated in more depth in [132]. King et al. [244] found differences in preference for listening over headphones versus loudspeakers, but these differences were not consistent for the three stimuli that were used. They did not in general find significant differences for differing acoustic conditions (though post-hoc analysis suggested that the test subject's preference for acoustic conditions had a slight effect on level settings) [271].

It's also worth noting that balance fluctuates over time within the course of a single trial, as the focus shifts between sources or new sources are introduced or removed. In [273] and related papers, Allan and Berg performed a series of experiments where engineers adjusted audio levels on authentic broadcast material as if running a live broadcast show. The aim was primarily to evaluate performance of, and human interaction with, real-time loudness meters. They found that the fader settings depended to a small extent on the use of different loudness meters and whether the test subject was a student or a professional. However, these experiments were concerned only with continually adjusting the overall level (riding the fader) to ensure consistency as program elements change, and did not deal with multiple simultaneous fader adjustments. Furthermore, there were differences in the experimental set-up between professionals and students, which may account for most differences between how they set the fader level.

Results of all of these experiments are dependent on statistical variations in the number of subjects, choice of subjects, and choice and range of stimuli. There may not be a hard and fast rule, and there may be further dependence on genre of the music, cultural background of participants and many other factors. Nevertheless, general patterns emerged across all studies, and any differences in level settings tended to stay within a few dB.

7.1.2 Intelligent Multitrack Faders

The most common form of multitrack automatic mixing system is based around simple level adjustments on each track. In almost all cases, it begins with the assumption that each track is meant to be heard at roughly equal loudness levels.

Dugan's work and subsequent automatic microphone mixers, described in Section 1.4.2, represent early attempts to automate multitrack faders. However, these methods are intended for speech signals and conference situations, where only one source should be heard at a time.

Excluding those early microphone mixers, Perez Gonzalez and Reiss [39] proposed perhaps the first intelligent music production tool in this area. It aimed to achieve ideal mixing levels by optimizing the ratios between the loudness of each individual input track, as well as the overall loudness contained in a stereo mix. To obtain real-time estimates of the loudness of each track, accumulative statistical measurements were performed. It then used a cross-adaptive algorithm to map the loudness indicators to the channel gain values.

Mansbridge et al. [46] provided a real-time system, using ITU-R BS.1770 as the loudness model [250]. An offline system described by Ward et al. [50] attempted to control faders with auditory models of loudness and partial loudness. In theory, this approach should be more aligned with perception, and take into account masking, at the expense of computational efficiency. But Wichern et al. [64] showed that the use of an auditory model offered little improvement over simple single band, energy-based approaches. Interestingly, the evaluation in [46] showed that autonomous faders could compete with manual approaches by professionals, and test subjects gave the autonomous system highly consistent ratings, regardless of the song (and its genre and instrumentation) used for testing. This suggests that the equal loudness rule is broadly applicable, whereas preference for decisions in manual mixes differs widely dependent on content.

7.2 Stereo Panning

7.2.1 Panning Assumptions

Panning (from 'panorama'), or the act of varying the gain between different channels, is an important but very simple mixing process. In a stereo setting, the pan position of a source is determined by a single parameter only, i.e., the relative balance between the left and right channel. Panning brings width to a recording, and makes sources more audible by spatially separating them.

Perhaps the most widely accepted mixing rule is that sources with a large amount of low-frequency energy should be panned center [12, 74, 120]. This consensus is at least partly informed by technical reasons. The directivity of both loudspeakers and human hearing is less pronounced at lower frequencies, and room acoustics aggravate this by typically reflecting low frequencies more strongly. As such, it makes little sense for low-frequency components to use up headroom of just one channel, especially since they are relatively higher in amplitude than high-frequency components. Another practical motivation for the centering of low-frequency energy is to reduce the chance that a vinyl record player's needle jumps out of the groove, which is more likely when only one channel contains large amplitude variations.

On the other hand, in the early days of stereo, and more recent examples which seek to mimic this effect, drums and other instruments were panned so they existed in only one of the two channels. The reason for this is historical too, as some early stereo mixing consoles offered only 'hard' (i.e., 100%) left and right panning, and sometimes center (50% left, 50% right), though music producers may choose to adopt this approach for aesthetic effect.

Equally ubiquitous is the tendency to pan the lead vocal centrally as well [12]. It is usually the single most important component of a song, and the lead singer is often at center-stage. And vocals should be heard well even if standing close to just one speaker, or when one channel is unavailable.

Finally, the snare drum is another instrument which is typically centered [12], as it tends to provide the 'backbeat' which is a key part of the song's foundation. Another reason for this panning is the tendency to match a source's panning to its position in stereo microphone pairs containing the same source.

To improve intelligibility of each centered source, they are often panned slightly off-center not to spatially overlap. With a center that's already so crowded, the remaining sources have to be panned left and right to provide a sense of width or envelopment [12], and to further enhance the audibility of the sources. This means that other than the lead solo instrument (at any point in time), low-frequency sources, and snare drums, sources are typically panned so that the spectrum of the left and right channels are somewhat balanced [12], with higher frequency sources panned progressively towards extremes with increasing frequency [12, 49, 120]. However, quantitative analysis of mixes has shown that this relationship tapers off after the low mids, with low sources centred but mid to high frequency sourcces being panned equally wide on average [74]. The complete mix, then, should not be too narrow, but still have a strong central component [178].

Best practices for surround panning and more immersive spatial audio formats are very ill-defined [12], and an active area of research.

7.2.2 Intelligent Multitrack Panning

In one of the earliest automatic mixing papers, Perez Gonzalez and Reiss described an autonomous panning system based on several best practices in music production [36]. This work was extended in [43], which is also notable for being the first automatic mixing paper with subjective evaluation.

The premise of [49] is that one of the primary goals of stereo panning is to 'fill out' the stereo field and reduce masking. In this paper, Mansbridge et al. set target criteria of source balancing (equal numbering and symmetric positioning of sources on either side of the stereo field), spatial balancing (uniform distribution of levels) and spectral balancing (uniform distribution of content within each frequency band). They further assumed that the higher the frequency content of a source, the more it will be panned, and that no hard panning will be applied. Finally, it used a multitude of techniques to position the sources: amplitude panning, timing differences and double-tracking.

Pestana and Reiss [58] took a different approach, where different frequency bands of each multitrack are assigned different spatial positions in the mix. This approach is unique amongst intelligent multitrack mixing tools since it does not emulate, even approximately, what might be performed by a practitioner. That is, practitioners aim for a single position (albeit sometimes diffuse) of each source. However, it captures the spirit of many practical approaches since it greatly reduces masking and makes effective use of the entire stereo field.

In evaluating different configurations of a full automatic mixing system, Matz et al. [63] showed that dynamic spectral panning had a larger effect in the overall improvement provided by automatic mixing than any of the other tools they considered (intelligent distortion, autonomous faders and multitrack EQ).

7.3 Equalization

7.3.1 EQ Assumptions

Pedro Pestana [119] found that engineers perform equalization quite liberally in order to, among many other reasons, reduce masking and reduce salient frequencies. In particular, close attention is paid to avoiding masking of the frequency content of vocal tracks.

Equalization is fairly well understood as a practical tool in mixing. Two common and related techniques are 'mirror equalization' and 'frequency spectrum sharing' [127, 154]. Mirrored EQ involves boosting one track at a particular frequency and cutting the other track at the same frequency [98,127]. Frequency spectrum sharing involves carving out a portion of the frequency space for each track. In a simple two-track scenario, this could involve low-pass filtering one track and high-pass filtering the other.

Through subjective testing, Wakefield and Dewey [154] found a preference for boosting the masked track or for full mirror EQ, over just cutting the masker. However, these tests were limited to just two track scenarios, with limitations on EQ settings.

With regard to frequency spectrum sharing, the spectrum for each track can be divided into essential and nonessential frequency regions. The essential regions are most likely the highest amplitude portions of spectrum, and nonessential regions are most likely the frequency regions that are easy to attenuate with low impact on timbre change and loudness balance between the tracks. For a given track, the frequency regions that are mainly covered by other tracks can be attenuated [61,108]. This frequency spectrum sharing was investigated in [154], but only with high or low-pass filters on each of two tracks, instead of fully carving out the spectrum for a source. Thus, their results say little about this technique.

Pestana [12] found that subtractive EQ (cuts) are generally performed with a higher quality factor than additive EQ (boosts). This suggests that mixing engineers often use equalization to remove a particularly problematic and narrow frequency, such as one due to a room resonance or electrical hum. However, it is also often assumed that it is more reliable to attenuate the masked frequency regions instead of boosting the unmasked frequency regions [127,129]. This assumption is partly due to the risk of filter instability with improper use of a boost. Yet no evidence has been found to suggest expert mixers might tend to cut more than boost [12]. A tendency to favor boosts or cuts would impact overall loudness and the balance between the loudness of tracks. Also, mirror equalization allows one to focus just on the most problematic portion of the spectrum in the most problematic tracks.

A number of studies show that a target frequency contour exists for the complete mix [274,275]. This contour is similar to that of pink noise, with an additional roll-off in the low and high end.

7.3.2 Intelligent Single Track Equalizers

Until recently, little was done to attempt self-equalization of musical signals. The only notable example of research in automatic equalization was by Reed [9], in which an offline machine learning approach was used. Reed's machine needs to be trained manually by humans, after which it equalizes using nearest neighbor techniques.

So-called match EQs are already ubiquitous, dynamically filtering an input source to match a predefined spectrum in a smooth and perceptually pleasing way. This target is typically derived from a reference signal. Ma et al. [52] described an intelligent equalization tool that, in real time, equalized an incoming audio stream towards a target frequency spectrum. The target spectrum of a complete mix was derived from analysis of 50 years of commercially successful recordings [275]. The equalization curve was thus determined by the difference in spectrum between an incoming signal and this target spectrum. This concept can be applied to a per-track EQ, assuming target spectra from analysis of typical spectra of processed instruments [74].

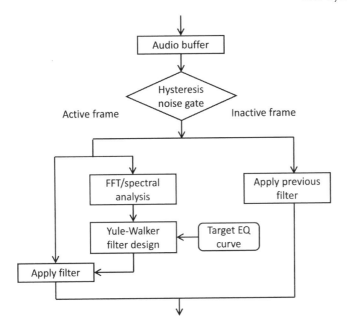

Figure 7.1 Block diagram of a real-time target equalization system

In this system, a hysteresis gate is first applied on the incoming signal, to ensure that only active content (i.e., not silence or low-level noise) is used to estimate the input signal spectrum. Since the input signal to be equalized is continually changing, the desired magnitude response of the target filter is also changing (though the target output spectrum remains the same). Thus, smoothing is applied from frame to frame on the desired magnitude response and on the applied filter. Targeting is then achieved using the Yule-Walker method, which can be used to design an IIR filter with a desired magnitude response. Figure 7.1 depicts a block diagram of this system.

Various approaches for learning a listener's preferences for an equalization curve with a small number of frequency bands have been applied to research in the configuration of hearing aids [276, 277] and cochlear implants [278], and the modified simplex procedure [279, 280] is now an established approach for selecting hearing aid frequency responses.

7.3.3 Intelligent Multitrack Equalizers

Perez Gonzalez [40] described a method for automatically equalizing a multitrack mixture. The method aimed to achieve equal average perceptual loudness on all frequencies amongst all tracks within a multitrack audio stream. Accumulative spectral decomposition techniques were used together with cross-adaptive audio effects to apply graphic equalizer settings to each track. Analysis demonstrated that this automatic equalization method was able to achieve a well-balanced and equalized final mix.

An alternative approach to autonomous multitrack equalization was provided by Hafezi and Reiss [61]. They created a multitrack intelligent equalizer that used a crude measure of auditory masking and rules based on best practices from the literature to apply, in real time, different multiband equalization curves to each track. The method is intended as a component of an automatic mixing system that applies equalization as it might be applied manually as part of the mixing process. Results of objective and subjective evaluation were mixed and showed lots of room for improvement, but also indicated that masking was reduced and the resultant mixes were preferred over amateur, manual mixes.

7.4 Dynamic Range Compression

7.4.1 Dynamic Range Compression Assumptions

Many assumptions regarding the use and setting of a dynamic range compressor were summarized by Ma et al. in [62]. They further performed a method of adjustment experiment to establish how audio engineers set the ratio and threshold. In a survey on the reasons to apply compression [12, 119], most professional mixing engineers who participated stated that their main intention for applying dynamic range compression (DRC) was to "stabilize erratic loudness range." They often compress instruments that have high note-to-note level variations, such as vocals or drum tracks, so that their relative levels are more consistent. Based on analysis of mixes, Pestana showed that "Compression takes place whenever headroom is at stake and the low-end is usually more critical" [12, 119]. Thus, a signal with more low-frequency content would typically have more compression. Spectral features of the source audio signal such as spectral centroid, spectral spread, and brightness are worth exploring to reveal the degree of frequency dependence and low-end sensitivity of DRC. Furthermore, the amount of typical compression depends on the instrument, more than the features [12], with the exception of spectral features.

Mix bus compression resulting in close to 3 dB RMS level reduction appears to be customary, and limiting peaks of up to 6 dB are imperceptible to even professional users [12]. However, compressing mostly on a per-track basis may reduce the risk of pumping and excessive distortion, and could therefore be preferable in many cases. But the evidence for this is not yet well-established, since the only study that investigated this aspect used unusual compressor settings and only one song as the stimulus [281].

Subjective comparison of loudness equalized mixes of the same song reveal a preference towards high dynamic range (typically achieved by less compression, if not louder transient instruments like drums) [178], suggesting it is better to err on the 'light' side of compression [48, 136, 281, 282].

A number of dynamics features have been proposed recently that measure the degree of level fluctuation, including EBU loudness range [250] and dynamic spread [22], which is simply the p-norm of the signal. Yet subjective testing in [283] suggested none of the metrics accurately predict the perceived dynamic range of a musical track. Pestana et al. [225] proposed parameter alterations to [250] that might yield better results for multitrack material. Alternatively, the crest factor, calculated as the peak amplitude of an audio waveform divided by its RMS value, can also be a coarse measurement of dynamic range.

It is generally accepted that attack and release time parameters are used to catch the transients in a signal, and hence their settings depend on the transient nature of the signal [108, 284]. Attack times usually span between 5 ms and 250 ms, and release times are often within the 5 ms to 3 s range. Some commercial compressors offer a switchable auto-attack or auto-release, which is mostly based on measuring the difference between the peak and RMS levels of the side-chain signal. McNally [20] proposed to scale the release time based on RMS values, whereas Schneider and Hanson [21] scaled attack and release times based on the crest factor. More recently, Giannoulis et al. [53] revisited the subject and scaled both parameters based on either modified crest factor or modified spectral flux, subsequently used by Maddams et al. [48]. The available body of research shares a general idea: if a signal is highly transient or percussive, shorter time constants are preferred.

A soft knee enables a smoother transition between non-compressed and compressed parts of the signal, yielding a more transparent compression effect. In order to produce a natural compression effect, the knee width should be configured based on the estimated amount of compression applied on the signal. This amount of compression largely depends on the relationship between threshold and ratio.

Make-up gain is usually set so that output loudness equals input loudness, making up for the attenuation of louder components, as the name suggests. Automatic make-up gain based on the average control-voltage is commonly used in commercial DRC products. However, in practice, this often leaves a perceived loudness variation [108]. Subjective evaluation showed that the EBU loudness-based make-up gain produced a better approximation of how professional mixing engineers would set the make-up gain [53].

Quantitative descriptions regarding the amount of compression that should be applied on different instruments can be found in [285, 286]. Giannoulis et al. [53] performed a discrete separation between transient and steady state signals and allowed the former to have larger variability of the knee width in order to accommodate for transient peaks. Ronan et al. [287] found that at least some mixing engineers apply separate compression to percussive subgroups, since they are assumed to need a different treatment from sustained sounds.

7.4.2 Intelligent Single Track Compressors

Due to its side-chain mechanism, the compressor is an early adaptive audio effect. So it is natural that expanding the side-chain's functionality would also be one of the first auto-adaptive strategies to arise. In fact, other than gain control, this is the earliest processing type to be rendered automatic. Nevertheless, dynamic range compression is more challenging to automate than other effects, as it is a nonlinear effect with feedback, there are complicated relationships between its parameters, and its use is generally less understood than other effects.

A reasonable approach to automation of the attack time would be to make its value a function of the reason to compress. That would suggest it depends on the dynamics or frequency range of the source, and hence could use the value of features such as loudness range, crest factor or spectral centroid. And louder events suggest faster attacks, so the attack time could be a function of short-term loudness. Pestana explored the validity of these approaches in [119] .

Tyler [19] describes a one-parameter compressor, where the input signal controls not only the gain, but also the soft knee's width. Other studies focus on automatic control of compressor time constants based on RMS level [20] and crest factor [21], and more recently, these were replaced by a modified crest factor and modified spectral flux [48]. In [53], Giannoulis et al. further modified the auto-adaptive nature of the compressor so that only one parameter, the compression threshold, was needed to control the unit. Finally, this was taken one step further by Mason et al. [60], who based the amount of applied dynamic range compression on a measurement of the background noise level in the environment.

The target curve approach (see the 'match EQs' in Section 7.3.2) can also be used for dynamics matching, and is exceedingly relevant in mastering. The parameters on a dynamic range compressor are set in such a way that the 'dynamics profile' – a cumulative relative frequency plot of frame loudness values – approaches the target curve [22, 59]. Dynamic properties of music signals have been investigated in detail [288–291], and they have also been measured in hit recordings over several decades [115, 292, 293]. Dynamics profiling and matching has been further explored in [75, 294] and others. In [59], for instance, Hilsamer and Herzog use the statistical moments of a signal's probability density function to determine offline dynamics processing for mastering applications.

To estimate the settings of a compressor with hidden parameter values, Bitzer et al. proposed the use of purposefully designed input signals [295]. Although this provides insight into predicting DRC parameters, in most cases, only the audio signal is available and the original compressor and its settings are not available for black-box testing.

At this time, the only machine learning approaches toward compression are in [66] and [296]. In [66], Mimilakis et al. emulate DRC for mastering using a deep neural network, trained on a set of songs before and after mastering. The use of deep neural networks for audio effects is quite novel, although only intelligent control of the ratio factor was considered.

Sheng and Fazekas [296] proposed the use of sound examples to control effects. Their focus was on estimating the parameters of a compressor given a sound example, such that the processed audio sounds similar in some relevant perceptual attributes (e.g., timbre or dynamics) to the reference sound. They used a set of acoustic features to capture important characteristics of sound examples and evaluate different regression models that map these features to effect control parameters. This requires the prediction of each parameter. Their approach consists of an audio feature extractor that generates features corresponding to each parameter, a regression model that maps audio features to audio effect parameters and a music similarity measure to compare the processed and the reference audio. As they only used simple monotimbral notes and loops as audio material, however, the applicability of their approach to musical material remains an open question.

7.4.3 Intelligent Multitrack Compressors

A first attempt at multitrack dynamic range compression was provided by Maddams et al. [48]. The perceptual signal features *loudness* and *loudness range* were used to cross-adaptively control each compressor. The system included six different modes of fully autonomous operation. Results of evaluation were mixed, and it was difficult to identify a preference between an automatic mix, a manual mix, and no compression applied at all. Furthermore, it wasn't possible to tell whether this was due to a genuine lack of preference or due to limitations in the experimental design (e.g., poor stimuli, untrained test subjects).

A more rigorous approach was taken by Ma et al. in [62]. The challenge was to formalize and quantify the relevant best practices described in [12]. First, a method of adjustment test was performed to establish preferred parameter settings for a wide variety of content. Then least squares regression was used to identify the best combination of candidate features that

Table 7.2 Studies concerning perception of reverberation of musical signals, from [82]. Test method: PE or DA (Perceptual Evaluation or Direct Adjustment of reverb settings); participants Skilled or Unskilled in audio engineering. Reverberator properties: Stereo or Mono; Early Reflections or No Early Reflections

		Stereo		Mono	
		ER	No ER	ER	No ER
PE	Skilled	[12, 298, 308, 313]	/////////	[302, 312]	[303]
	Unskilled	[299, 304–307]	[301]	[253]	[252, 300]
DA	Skilled	[12]	[254]	[309]	[310]
	Unskilled	[306, 311]	/////////////////////////		[252]

map to parameter settings. Thus a rule such as "more compression is applied to percussive tracks" translates to "the ratio setting of the compressor is a particular function of a certain measure of percussivity in the input audio track." Perceptual evaluation showed a clear preference for this system over amateur application and over no compression, and sometimes a similar performance to professionals.

7.5 Reverb

7.5.1 Reverb Assumptions

Of all the standard audio effects found on a mixing console or as built-in algorithms in a digital audio workstation, there has perhaps been the least effort on furthering understanding of reverberation [12, 82, 105, 119]. This may seem counterintuitive, as reverberation is an important factor towards the perceived quality of a mix [82], and as a linear, time-invariant phenomenon it should therefore be relatively straightforward to analyze [297]. However, there is a notable lack of universal parameters and interfaces, and a large variation of algorithms across reverberators [82, 119]. In comparison, typical equalization parameters are standardized and readily translate to other implementations. A number of studies have looked more generally at perception of reverberation in musical contexts [252–254, 298–313] (see Table 7.2).

Reverberation is characterized by a large number of perceptual or physical attributes. Most work thus far has been done on the notion of perceived reverb amount, which correlates positively with at least objective loudness [12, 82, 253, 301], decay time [82, 298, 301, 305, 307, 308, 313], and density [119]. However, these relationships are nonlinear [253].

Practical audio engineering literature offers few actionable reverb settings or strategies [54], perhaps again due to the high diversity and complexity of interfaces and implementations. Interviews with experts show reverb signals are best filtered with a 200 Hz low-pass filter and 5 kHz high-pass filter before or after processing, though this practice and these values are far from agreed upon [119].

In a multitrack mixture, the perceived reverberation is the result of different reverb types applied to several sources to various degrees. Furthermore, the resulting tracks and sums thereof are subsequently affected by other processing – some of it nonlinear. As such, multitrack reverberation cannot adequately be described by an impulse response, as opposed to the acoustic reverberation between two points in a room. To address this, De Man et al. quantified the perceived amount of reverberation from a multitrack mix in terms of the relative reverb loudness and the equivalent decay time of the combined reverberation [82]. Analysis of these quantities over 80 full mixes from 10 songs revealed these features were indicative of which mixes were subjectively considered to have too much or too little reverberation. It also zoned in on −14 LU as a suggested middle ground for reverb loudness (if the total mix is 0 LU). This is in contrast with [12], where six songs with reverb at six artificially adjusted levels measured the highest preference for the −9 LU difference. There seems to at least be a consensus that it is better to err on 'dry' side [12, 82, 230], a caution which can be incorporated in automatic reverb implementations.

The desired or expected reverberation amount depends on the playback conditions. For instance, skilled listeners preferred higher levels of reverberation when using loudspeakers than when using headphones [134] and when in a relatively less reverberant space [133] – based on only one test stimulus – and [253] measured an inverse relation between perceived reverberation amount and playback level.

A number of studies report a possible link between ideal reverberation time and the auto-correlation function of the (dry) signal [12, 314]. The decay would also be inversely correlated with spectral flux, and effect which is even stronger when applying a logarithmic transformation to both features.

Furthermore, it has been suggested that reverb brightness should be higher for long reverb times and dull sounds; high fidelity reverbs pair well with 'trashy' sounds and vice versa; high spectral flux indicates a need for short reverb times; sparse and slow arrangements require more reverb; percussive instruments require shorter and denser reverbs than sustained sounds; and low-frequency sounds require less [119].

The most esoteric reverb parameter investigated thus far is the predelay (the lag between a dry sound and its first simulated reflection), which can be set just past the Haas zone, i.e., the time window of 30–50 ms within which the ear perceives several sonic events as one sound [119].

Related to and sometimes used instead of reverberation is the delay effect, where one or more delayed and sometimes processed versions of the signal are added at decreasing levels. A delay, or echo, can be defined as a discrete repetition of the signal after a fixed period of time, typically above 30 ms (because of the aforementioned Haas effect) [105]. Because of this, it is perceived as separate from the dry signal, and does not significantly alter the timbre through comb filtering. Substituting reverb with a delay effect can be advantageous to reduce masking [12]. Pestana et al. [105] found a preference for delay times coupled with song tempo, where eighth-note delays (twice per beat or eight times per bar) on a vocal work better than longer ones.

7.5.2 Intelligent Single Track Reverb

An early implementation of reverberation effect automation was proposed by Heise et al. [297], who optimized parameters of off-the-shelf reverberation effects so that the resulting reverb approximates a given impulse response. The choice of impulse response and reverb level is still up to the user, but it adds the benefits of low computational complexity and ease of adjustment to the use of IR-based reverbs, which can be based on real spaces. Its performance was measured to be on par with or exceeding that of professional human operators.

In 2016, Chourdakis and Reiss [65] created an adaptive digital audio effect for artificial reverberation that learns from the user in a supervised way. They first performed feature selection and dimensionality reduction on features extracted from a training data set. Examples of reverberation parameters are user-generated for the training data. Finally, a set of classifiers is trained, which are compared using tenfold cross-validation to compare classification success ratios and mean squared errors. Tracks from the Open Multitrack Testbed [315] were used in order to train and test the models (see Section 5.1.2). In a follow-up paper, mappings from effect parameters to standard reverberation characteristics from [265] were incorporated in this model, and the results evaluated subjectively [69].

Benito and Reiss [70] presented an approach based on Probabilistic Soft Logic (PSL) rules, derived from results of the aforementioned studies [82, 105, 119] and best practices recommended by experts in practical audio engineering literature [108, 118]. The use of PSL importantly allows prediction of parameters with varying levels of confidence. Features were extracted from the incoming tracks to determine the likely instrument and some high-level descriptors (*bright*, *percussive*, *voiced*), which helped determine which rules to follow. The system was presented without quantitative or perceptual evaluation.

One important aspect of reverb that is yet to be addressed in automation is spatial reverb. In order for a listener to perceive the 'spaciousness' of a large room, the sounds that reach each of the listener's ears should be slightly decorrelated. This is one reason why concert halls have such high ceilings. With a low ceiling, the first reflections to reach the listener usually are reflections from the ceiling, and reach both ears at the same time. But if the ceiling is high, the first reflections come from the walls of the concert hall. And since the walls are generally different distances away a listener, the sound arriving at each ear is different. Ensuring some slight differences in the reverberation applied to left and right channels is important for stereo reverb design.

7.6 Distortion

7.6.1 Distortion Amount Automation

Because it is such a creative effect, no best practices are available on which complete automatic parameter setting can be based. However, some approaches for automating aspects of distortion effect control have been proposed in [56].

As the amount of distortion (often dubbed 'drive' or 'overdrive') usually increases with increasing input signal level [110, 316], a static distortion device needs careful tuning of the gain or threshold controls to apply the right amount of distortion. This also means that static settings inevitably introduce varying amounts of distortion for different input signal levels. This may be appropriate in some cases – after all, this is the expected behavior of the intrinsic saturation induced by analog devices such as amplifiers – but can be detrimental in situations where a constant degree of distortion is desired, leaving soft parts virtually unprocessed or louder parts heavily damaged. It further requires substantial time and effort to dynamically tailor the characteristic threshold of the distortion input-output curve to the levels of new input sources.

De Man and Reiss [56] presented a distortion effect that scales the input-output characteristic automatically and in real time proportional to a moving average of the signal's level. Instead of a low-level parameter such as 'threshold' or 'input gain,' where knowledge of the signal is required to predict the amount of distortion, the parameter introduced here is closer to the semantic concept of 'amount of distortion.'

It would be undesirable to control the amount of distortion by an input gain parameter, which increases the effective level and, in the case of multiband distortion, changes the relative level of the different bands. Rather, parameters of the device's input-output characteristic are controlled, manipulating for instance its clipping point, while compensating automatically for the loss in loudness or energy induced by clipping the signal's peaks. Another reason for scaling the input-output characteristic rather than the input gain to control the amount of distortion is that the latter introduces a larger level difference when the effect is switched on, making fair A/B comparison impossible.

In order to obtain an equal amount of distortion (for a given input-output curve) for signals of varying level, an 'auto-threshold' function is implemented, similar to the automatic threshold in automatic compressor designs [48]. Instead of threshold T, the user then has control over the the relative threshold T_{rel}, which relates to the threshold as follows:

$$T = T_{rel} \cdot L_{RMS} \tag{7.1}$$

where L_{RMS} is the moving RMS level and T_{rel} is the threshold relative to the smoothed RMS level. Unlike the effective threshold T, this relative threshold is not necessarily less than 1.

To ensure that the same amount of distortion is applied for different levels, the threshold is scaled with the square root of the exponential moving average (EMA) of the square of the signal value (moving RMS):

$$L_{RMS}[n] = \sqrt{(1 - \alpha) \cdot x^2[n] + \alpha \cdot L^2_{RMS}[n-1]} \tag{7.2}$$

where $x[n]$ is the value of sample n, and

$$\alpha = \exp\left(\frac{-1}{\tau \cdot f_s}\right) \tag{7.3}$$

where τ is the characteristic time constant in seconds, and f_s is the sampling rate [90]. This means that, by definition, changing the signal level by a certain amount changes the threshold by that same amount.

A benefit of this method is that the dynamic range is left unharmed, even for very high degrees of distortion. As an example, static hard clipping with a very low threshold could bring down the level of a rather loud fragment to the level of a subsequent soft part of the same source, as shown in Figure 7.2(b). When made adaptive as above, however, the long-term dynamics will be maintained as not only the same distortion will be applied, but the relative level differences will also be maintained. This is shown in Figure 7.2(c). Depending on the value of the time constant, relatively short-term dynamics could still be tempered.

As evident from Figure 7.2(c), the lag inevitably introduced by the exponential moving average causes the threshold to react late to onsets of parts with a significantly different level. Consequently, a loud signal immediately following a quiet one would temporarily be distorted more than desired, and vice versa. This can be compensated by choosing ever lower time constants, causing the threshold, and hence the amount of distortion introduced, to change much faster. Another solution to this problem is the introduction

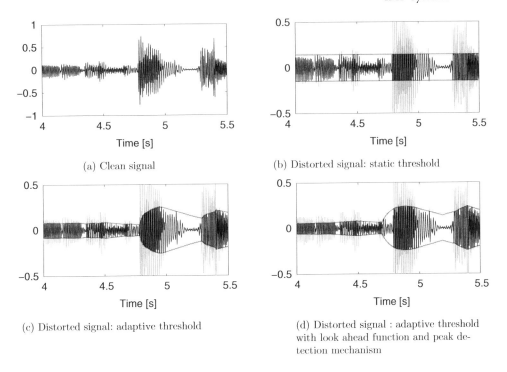

(a) Clean signal

(b) Distorted signal: static threshold

(c) Distorted signal: adaptive threshold

(d) Distorted signal : adaptive threshold with look ahead function and peak detection mechanism

Figure 7.2 Fragment of bass guitar recording with sudden level change. In (b) through (d), the light gray signal is the original, clean signal, while the black signal is the resulting signal after symmetric hard clipping distortion with the threshold shown in dark gray.

of a lookahead function (delaying the output while the input is considered K samples in advance) in combination with peak detection into the previously described, sample-based system. Equation 7.2 then becomes:

$$L_{rms}[n] = \sqrt{\begin{array}{l} (1 - \alpha) \cdot \max\left(x^2[n], x^2[n+1], ..., x^2[n+K]\right) \\ + \ \alpha \cdot L_{rms}^2[n-1] \end{array}} \quad (7.4)$$

The effect of using such a lookahead buffer (with characteristic adaptive time constant $\tau = 200$ ms, the corresponding EMA coefficient α, and lookahead buffer size $K \cdot f_s = 100$ ms) is shown in 7.2(d).

Virtually any amplitude distortion input-output characteristic is suitable for threshold automation. The underlying principle is to scale the input variable x with $1/L_{RMS}$ to ensure that similar signals with different levels (captured in L_{RMS} for sufficiently long-term dynamics) have the same relative amount of distortion, and to multiply L_{RMS} with the output to preserve dynamics. The input-output characteristic should be scaled with the relative threshold, T_{rel}, to determine the amount of distortion.

However, some elementary nonlinear input-output characteristics, like the full-wave or half-wave rectifier, require no such scaling as the induced distortion does not depend on signal level. Rectifiers are known to introduce even harmonics (as opposed to e.g., symmetric

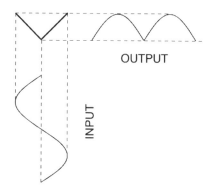

Figure 7.3 The effect of an ideal rectifier on a sinusoidal input

hard clipping, which introduces only odd harmonics), to an often pleasing sonic effect. An ideal rectifier simply behaves as follows:

$$f(x) = |x| \tag{7.5}$$

in which case the fundamental of a sine wave disappears completely (see Figure 7.3). A half-wave rectifier is equivalent to equal parts unprocessed and fully rectified signal.

Matz et al. tested an 'automatic exciter' based on this threshold automation method in [63], which improved the audibility, assertiveness, presence, clarity and brightness of the source, by adding harmonic distortion components to the high frequency range. While not formally tested separately, as every stimulus is the product of a combination of processing blocks and the authors chose to use the automatic exciter in each of these, several of the systems tested indeed show an improvement in transparency, timbre and overall impression.

7.6.2 Subjective Evaluation

A listening study with 12 participants was conducted to evaluate adaptive distortion with and without look-ahead, alongside a clean (reference) and statically distorted sample (anchor). The results revealed a high perceptual consistency throughout each sample for the adaptive distortion case. Similarly, the difference in perceived distortion level between the low and high volume part of the fragment was close to zero on average, although with significantly greater variance than for the clean reference. No difference in performance was measured between adaptive distortion with and without lookahead.

In 33 out of 36 responses, the static distortion was found to be less consistent in terms of perceived distortion than either of the adaptive ones. Again in 33 of 36 responses (though not all the same ones), the perceived difference in terms of distortion between the loud and soft part was rated higher in the static case than it was in the adaptive cases.

Although the degree of nonlinearity (level of harmonics) is mathematically the same regardless of the level after a transition period, there was a slight tendency to rate the louder part as more distorted in the adaptive cases.

7.6.3 Adaptive Anti-aliasing

Like other nonlinear effects, amplitude distortion introduces harmonics at frequencies that potentially were not present in the original signal. This means it may be necessary to upsample before applying distortion in order to avoid audible aliasing artifacts [317]. The extra frequency components introduced by aliasing are almost always inharmonic, and therefore deteriorate the perceived sound quality [113]. For this reason, an interpolation filter is applied to the signal before applying any kind of processing that introduces new harmonics. Afterwards, if an audio output at the original sampling frequency is desired, a decimation filter should be applied. Naturally, a higher upsampling ratio is more computationally expensive.

Knowing the signal's or band's approximate upper frequency limit and estimating the number of harmonics that should be below the Nyquist frequency, an adaptive upsampling ratio can be found to save computational power while ensuring a sufficiently low perceptual difference. For a signal where the highest significant signal component is f, the desired number of harmonics (including fundamental) to fall below the new Nyquist frequency is N, and the original sampling rate is f_s, the necessary upsampling ratio can be found as follows:

$$R = \left\lceil \frac{2Nf}{f_s} \right\rceil \tag{7.6}$$

Further automation of the upsampling ratios could be achieved by making them dependent on the type of distortion (more specifically the relative levels of the respective harmonics), user input (the desired maximum amount of computation and/or minimum degree of quality) and possibly the time-varying spectrum of the input signal. For polynomial input-output curves of nth order, new frequency content is only generated up to the nth harmonic [112], making even simpler, real-time and near-perfect adaptive upsampling possible [318].

7.6.4 Other Automation

To address the possible change in loudness induced by the distortion, another automation mechanism can be implemented. A loudness metric of choice is measured before distortion, and then again after applying distortion. The distorted signal is then multiplied with a smoothed version of the ratio of these values L_{in}/L_{out}. This can minimize the perceivable difference in loudness or level when the device is switched on or off, or when the wet/dry mix (relative strengths of distorted and undistorted signal in a combined output) takes another value. It also avoids a level drop of the distorted signal when the threshold is brought down, due to 'chopping off' the signal's peaks to a degree where little remains of the original signal. In the case of multiband distortion, it prevents the bands from changing their relative loudness (e.g., a distorted, louder high band), as the absolute loudness or level of every single band is maintained.

Finally, harmonic distortion introduces energy even at frequencies where the original signal had none (as opposed to linear effects like equalization and reverberation). Parameter automation offers the opportunity to maintain a specified frequency balance in ways other processors cannot. One can impose a certain ratio of the energy of the high band versus the energy of the rest of the signal, and mix in more or less high band distortion to achieve this target ratio and avoid dullness or piercing highs. In this case, the energy of a high-pass filtered signal above $2 \cdot f_c$ is compared to the total signal energy (where f_c is the cutoff frequency of the high-pass filter), and the distortion of the high-pass filtered signal x_{HPF} is balanced accordingly. This affects the energy of the frequencies above $2 \cdot f_c$, as harmonic distortion of the high band starting at f_c will have its second harmonic at $2 \cdot f_c$ or higher. This can be considered an automatic exciter, ensuring a consistent amount of brightness in the processed signal. This approach was taken by Matz et al. [63] as part of a comprehensive automatic mixing system.

Although many intrinsic distortion parameters can be automated through the above measures, there are still at least as many parameters that need tweaking (e.g., time constants τ and τ_L, relative threshold T_{rel}, the target high-to-low energy ratio). However, these measures can be set to appropriate values once, possibly saved as a preset, and are by definition independent of the signal level from that point onwards, making it safe to leave them unchanged in an autonomous mixing situation.

7.7 Instrument-specific Processing and Multi-effect Systems

Many of the intelligent mixing systems described in this chapter lack instrument-specific processing. Mixing decisions are based solely on the extracted, low-level features of the signals and no high-level semantic information, such as which instruments the incoming tracks accommodate or the genre of the song, is provided by the user or extracted by the system. Notable exceptions are those implementations where one or more principal tracks can be manually selected by the user to force a level increase [46] or fixed panning position [49].

Taking this concept further, De Man and Reiss [54] designed a fully knowledge-engineered system where rules depend on a manually assigned instrument label, and Scott and Kim [55] applied fixed equalization curves, fixed pan positions and an adaptive gain coefficient to drum tracks.

8

IMP Processes

"Human beings have dreams. Even dogs have dreams, but not you, you are just a machine. An imitation of life. Can a robot write a symphony? Can a robot turn a canvas into a beautiful masterpiece?"
—*"Can you?"*

Detective Del Spooner, portrayed by Will Smith, and robot Sonny, portrayed by Alan Tudyk, in *I, Robot*, 2004.

8.1 Fundamentals and Concepts

Chapter 7 looked at automation of audio effects in order to achieve standard audio production goals. The emphasis was on how to establish autonomous use of a specific audio effect within the context of multitrack mixing, e.g., setting equalizer parameters for each track in a mixture. However, many tasks in audio production are achieved by complex processes which require the use of several different audio effects, or may not even be achieved by the use of audio effects. For such tasks, the emphasis is more on the process of achieving the task, and less on choosing the appropriate parameter settings of the effects that are used.

Here, we have grouped together the approaches to such tasks as Intelligent Music Production *processes*. They are still tackled using advanced signal processing and machine learning methods, coupled with in-depth knowledge of best practices and psychoacoustics. But they are focused around questions like

"How do I get rid of unwanted noise?"
"Can a machine generate sequences that sound human?"
"In what order should effects be applied?"

"How can I estimate the parameter settings applied in an existing mix?"
"How should I mix for on-stage performers as opposed to front of house?"

Many of these tasks are more technical than they are artistic, and to some extent menial and tedious. As such, they can detract from the creative flow of the sound engineer or music producer, and are a prime candidate for automation or streamlining.

8.2 Editing

8.2.1 Automatic Take Selection

"If at first you don't succeed..."
A studio recording is rarely the result of a single, uninterrupted performance. In fact, the opportunity to iterate, experiment, and try again is among the most fundamental distinctions between the studio environment and a live setting. It is very common for an individual musician, or an ensemble, to record many repetitions of a song, section or even phrase. Despite the wide range of corrective processing available to the sound engineer, it is often more viable to try and capture a better version than to correct artificially. With every extra take, though, the task of judging takes or even parts of takes on their aesthetic and technical merits becomes more challenging.

Bönsel et al. [319] propose a system that automates this task, for the specific cases of selecting electric guitar and vocal takes. Disregarding the highly subjective aspects of dynamics and timbre, they rank performances based on 88 features relating to pitch with respect to the equal tempered scale, and timing with respect to a click track. In their evaluation, the take selection system performs well for guitars, and slightly less for vocals due to a higher rate of transcription errors.

8.2.2 Structural Segmentation

Despite everything digital technology has made possible, navigation in an audio workstation is still very 'analog,' where the playhead (note the tape-era term) has to be positioned manually by clicking, dragging, or holding down a button. The burden of recognizing or remembering where a certain section or event was, and creating markers as bookmarks throughout the session, is still on the user. Visualized waveforms offer only limited help, as they show level variations, and areas where different tracks are active, but not much else. This is in stark contrast to text processors, where a navigable table of contents or a word search facilitates browsing of even very large documents [174].

The most obvious and intuitive navigation functionality would allow a user to go to the start of a segment of the song's 'form,' that is, a chorus, verse, bridge, or other high-level structural entity. In the context of audio production, more so than in consumer-facing playback systems, access to multitrack audio can be leveraged to enhance the segmentation process [174,320]. Certain works have focused on the detection of instrumental solo intervals [320,321].

8.2.3 Time Alignment

In live settings, multiple acoustic sources are typically recorded with multiple microphones, close to the source they are intended to pick up, sometimes aided by a directivity pattern that attenuates sound from particular directions. Even so, it cannot be helped that each microphone will also to some extent pick up other sound sources. As a result, each microphone signal contains versions of most other sources, be it at a lower level and with a slightly higher delay due to acoustic propagation. This is true for studio recordings as well, when musicians perform at the same time while not being in isolated spaces.

When these microphone signals are summed, the differently delayed versions of the respective signals cause comb filtering. By artificially delaying all but one microphone to various degrees, unwanted coloration can be reduced. A related problem occurs when a particular source is recorded by several microphones at different distances. In that case, the closer microphones need to be delayed accordingly so that the recorded signals line up.

Delay and interference reduction are actually well-established signal processing techniques, and have been customized extensively for mixing applications [72, 148, 322–326]. These algorithms deal with optimizing parameter settings for real world scenarios, such as microphone placement around a drum kit, moving sources on stage, or by restricting interference so that no additional artifacts are introduced to a recording.

Perez Gonzalez and Reiss [72] approached the problem of time alignment and automation of track delays by approximating the transfer function between two signals. The inverse Fourier transform of the phase of that transfer function then gives an impulse response, and its maximum value corresponds to the delay between signals. In this approach, smoothing and other tricks were used to ensure robust behavior in difficult environments. However, the main concept is almost equivalent to the most common approach to automatic time delay estimation, known as the Generalized Cross-Correlation Phase Transform (GCC-PHAT). This method calculates the phase difference between the two signals, which is then transformed to the time domain to estimate the delay.

The Generalized Cross-Correlation, or GCC, is defined as:

$$\psi[k] = X_1^*[k] \cdot X_2[k] \tag{8.1}$$

where X_1 and X_2 are frequency domain representations of two signals x_1 and x_2, $k = 0, ..., N-1$ is the frequency bin number, and * denotes the complex conjugate. GCC-PHAT uses only the phase of the GCC, which has been shown to improve performance in noisy and reverberant conditions.

Equation 8.1 then becomes:

$$\psi_P[k] = X_1^*[k] \cdot X_2[k]/|X_1^*[k] \cdot X_2[k]| \tag{8.2}$$

The delay is then estimated as the value for ψ_P that maximizes the inverse Fourier transform, denoted here as F^{-1}.

$$\tau = \arg\max_n F^{-1}\{\psi_P[k]\} \tag{8.3}$$

Once the delay is found, it is a simple task to apply that delay to the earlier signal in order to time align any two signals which capture the same source. It can be easily extended to many tracks by treating one track as reference and calculating the GCC-PHAT for each of the other tracks with the reference.

In a number of studies, Clifford and Reiss [324, 325] focused on the challenges of using GCC-PHAT with musical signals, an important issue since most previous investigations of its performance were limited to speech or test signals. Their work uncovered often overlooked parameter settings and signal features which strongly influence the performance of the technique, and resulted in recommended implementations. Jillings et al. [326] further investigated optimal choices of parameter settings for a wide range of scenarios (moving sources, background noise, reverberation...) dealing with recording and mixing real world musical content. In an earlier study, Clifford showed how GCC-PHAT can also be used to calculate each of the time delays when there are at least as many sources as microphones [148]. However, in this situation, there may not be a preferred delay to apply, since aligning two microphone signals to one source would often cause misalignment with another source.

8.2.4 Polarity Correction

Similar to but distinct from the issue of delay compensation, is the correction of polarity differences. When tracks which are part of a mixture containing similar recordings of the same source, but with inverted polarity, their summation can result in some degree of cancellation due to phase interaction. A typical example is that of a snare drum being simultaneously recorded by a 'top' and 'bottom' microphone. The two captured signals are very similar to each other, but can be almost inverse. The common solution to this is flipping the phase of one of the recordings, so that summing them results in constructive rather than destructive interference. The GCC-PHAT algorithm described above can also be used to detect when signals do not have the appropriate polarity and should be inverted [326]. Selectively changing the polarity of sources in a mix can be used as a way to reduce the total peak level, thereby maximizing loudness [327]. Results show loudness increases of around 3 dB, while sounding more transparent and being less prone to artifacts than dynamic range compression or limiting.

8.2.5 Noise Gates

Noise Gate Assumptions

Noise gates eliminate noise which may otherwise be amplified and heard when an instrument is not being played, by intermittently attenuating or muting the entire track. In many cases, they are used to remove spill from other instruments, temporally isolating the sources to allow independent processing. For instance, a gate may be used on a snare drum when it needs brightening, without also brightening the cymbals captured by the snare drum microphone.

The choice of parameter settings in a noise gate is a delicate balance between reducing the level of bleed and preventing distortion of the source signal. The threshold should be set high enough that the ambient noise falls below it, but not so high that the instrument's decaying notes are cut off prematurely. If the threshold is set below the peak amplitude of any part of the bleed signal, then the bleed will open the gate and will be audible. If the release time is short, the gate will be tightly closed before the crosstalk from other sources, but the natural decay of any notes in the source will be cut off abruptly. If the release time is long the gate will remain partially open, and the spill will be audible to some extent, but the source will be allowed to decay more naturally.

When a noise gate is used in drum tracks, it is common practice to first set gain to $-\infty$ dB, i.e., to completely mute the 'bleed' [42]. The threshold is then set as low as possible to let the maximum amount of clean signal pass through without allowing the gate to be opened by the bleed signal. The release is set as slow as possible while ensuring that the gate is closed before the onset of any bleed notes. For very fast tempos this may not be possible without introducing significant artifacts, in which case some bleed may be allowed to pass through.

The attack is set to the fastest value that does not introduce any distortion artifacts. The hold time is continually adjusted to remove modulation artifacts caused by rapid opening and closing of the gate. During an inter-onset interval assigned to the clean source, the gate should go through one attack phase and one release phase only. The hold parameter should be as low as possible while maintaining this requirement. If it is too long it can affect the release phase of the gate. Once all other parameters have been set, the gain is adjusted subjectively to the desired level.

The noisy signal $y_n[n]$ is the sum of the clean source signal $y_c[n]$ and the bleed signal $y_b[n]$. Passing a noisy signal through the noise gate will generate a gate function $g[n]$ applied to the source. This vector contains the gain to be applied to each sample of the input signal. The gating will attenuate the source signal, generating distortion artifacts, D_a, and reduce the bleed to a residual level D_b

$$D_a = \frac{|(1 - \mathbf{g})^T \mathbf{y}_c|^2}{|\mathbf{y}_c|^2} \tag{8.4a}$$

$$D_b = \frac{|\mathbf{g}^T \mathbf{y}_b|^2}{|\mathbf{y}_b|^2} \tag{8.4b}$$

where we have used vector notation to represent the signals. The signal to artifact ratio (SAR) and reduction in bleed (δ_b) are then given by

$$SAR = 20\log_{10}(D_a^{-1}) \tag{8.5a}$$

$$\delta_b = 20\log_{10}(D_b) \tag{8.5b}$$

Intelligent Noise Gates

A traditional noise gate is an auto-adaptive effect, which determines whether or not to 'gate' the signal (reduce significantly in level or mute altogether) based on if its level is below the chosen threshold after smoothing with the chosen time constants. A more intelligent version of this processor would either set a number of these parameters automatically based on the properties of the signal itself, and possibly other signals, or determine which portions of the signal are undesirable using some other feature than instantaneous level.

A first implementation of an intelligent noise gate for drum recordings was described by Terrell and Reiss [328]. The clean signal is estimated from a recording of one clean note from the source, and then finding all frames within the noisy signal that are highly correlated with this recording. Choice of frame size and timings is highly dependent on detecting onsets, though many algorithms exist for this. Parameter settings are then found by minimizing a weighted combination of SAR and δ_b. Release and threshold are parameters in the objective function, but attack and gain are fixed at 1 ms and $-\infty$ dB respectively. Hold time is not applied. A usable automatic gate requires these parameters to be included. In particular, fixing the gain at $-\infty$ dB may cause abrupt attenuation of the decay of the clean source signal, even with a long release.

A more sophisticated implementation was later presented, which includes attack time and hold time as parameters in the objective function, and gain rather than a weighting function was used to control the strength of the gate [42]. Instead of minimizing an objective function which contains both distortion artifacts and noise, it attempts to optimize all parameter settings subject to minimizing the distortion while maintaining a minimum bleed reduction level.

Figure 8.1 is a block diagram of this algorithm. The inputs on the left are parameter values at each stage. The inputs on the right are constraints enforced at each stage. The signal is split into regions which contain just the clean signal and regions which contain bleed. An initial estimate of the threshold is found by maximizing the *SAR*, subject to the constraint that the bleed level is reduced by at least 60 dB. This is identified by the parameter δ_b, which is the minimum change in the bleed level after gating. The attack, release, and hold times are set to their minimum values during the initial threshold estimate and the gain is set for full signal attenuation. This ensures that the threshold is set to the lowest feasible value. The minimum hold time which results in exactly one attack and one release phase per onset window is then found. These constraints are identified by parameters $N_A = N_R = 1$ which correspond to the permitted number of attack and release phases, respectively, during a frame of the source signal. The other gate inputs are the minimum values of attack and release and the initial threshold estimate. The threshold estimate is required because the minimum hold time can vary significantly with threshold. The threshold is then recalculated using the updated hold parameter. Finally the attack and release are found by maximizing the *SAR*, subject to the bleed reduction. Gradient descent optimization methods are used to minimize functions at each stage.

The automatic microphone mixing systems described in Chapter 1 can also be considered early forms of intelligent gating based on relative level [24] or angle of incidence [27].

The removal of various types of noise is an extensive research topic in and of itself, and gating is only a small part of this. In many cases, the noise is active at the same time as the desired signal, which means that attenuating or muting the track in question altogether is not a viable option. Examples of offending signals are mains hum, wind noise in outdoor recordings, and acoustical reverberation. Reduction of the latter (dereverberation) is again an active research topic in its own right; others are usually denoted source separation.

8.2.6 Unmasking

There is no doubt that one of the main challenges in recording, mixing and mastering music is to reduce masking. This is discussed extensively in a number of technical guides for music production, whereby authors discuss the loss of clarity due to the presence of masking between sources [108, 119]. Alex Case [107] notes that "A simple rock tune [...] presents severe masking challenges... Guitars mask snare drums. Cymbals mask vocals. Bass guitar masks kick drum. An engineer must cleverly minimize unwanted masking."

But what do we mean? Music producers and music technology educators tend to use the term masking in an informal sense, often referring to sources "fighting to be heard." When there is a lot of masking, the mix is sometimes described as muddy or murky. Case defines masking as "When one signal competes with another, reducing the hearing system's ability to fully hear the desired sound" [107].

In psychoacoustics, we refer to masking as "The process by which the threshold of audibility for one sound is raised by the presence of another (masking) sound" [329]. Energetic masking occurs when the neural response of the combined masker plus desired signal is equal only to the individual neural response of the masker. That is, it is only due to how the auditory system processes the sounds. Informational masking occurs when there is difficulty in perceptually segregating the signal and the masker. Informational masking is essentially a catch-all for masking that cannot be explained solely by simple models of the auditory system. It covers higher-level factors, such as perceived similarity between simultaneous sounds [330], and includes linguistic and semantic context.

Further divisions are possible. There are two categories of auditory masking: simultaneous (spectral) masking and non-simultaneous (temporal) masking. Simultaneous masking refers to the case where both signal and masker are presented at the same time and primarily depends on the spectral relationship between the two sounds. Temporal masking may occur when both signal and masker do not overlap in time. In forward temporal masking, the masker occurs before the desired signal whereas in backward masking the masker occurs after the signal. Thinking of a loud bang preceding or following a spoken word, it is clear that forward masking is the stronger effect.

Finally, there is spatial release from masking or spatial unmasking. This suggests that there is a significant reduction in masking when the sounds come from different spatial locations, whereby the binaural nature of hearing can lead to a significant reduction in masking [331].

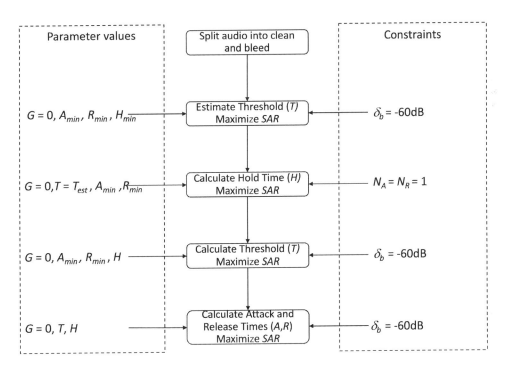

Figure 8.1 Block diagram of an intelligent noise gate

Wakefield and Dewey [154] show that avoidance of spatial masking may be a far more effective way to address general masking issues in a mix than approaches using equalizers, compressors and level balancing.

All these formal versions of masking deal with a change in the threshold for audibility, that is, the binary distinction as to whether a sound is heard or not. In practical situations, we are concerned with partial masking, where the loudness of the desired sound may be reduced but the sound is still perceived. For this, we make use of partial loudness, the loudness of a sound in the presence of another (masking) sound [269, 270, 332].

Based on this auditory model, masking in multitrack audio has been quantified with a masked-to-unmasked ratio [333], a cross-adaptive signal to masker ratio [334], measures of partial loudness in a mix [335] and problematic masking outliers [336].

Intelligent Unmasking Systems

Many of the automatic mixing tools presented in Chapter 7 use masking reduction as one of the goals of autonomous operation of a particular effect. For instance, Perez Gonzalez [38] used a form of 'mirror equalization,' which selectively unmasked a track by removing energy from competing frequency bands in other tracks. However, most of the methods referred to masking in an informal sense, without use of an auditory model. Even when an auditory model was used to estimate loudness, many of the systems did not use partial loudness or other formal approaches to estimating masking [39, 57]. Exceptions to this are [50, 64, 337], which incorporated perceptual masking measures into their rules for fader settings. Part of the reason for this is the computational complexity, and the lack of flexible real-time implementations of well-known auditory models [146, 338].

In practice, a multitude of different approaches to masking reduction are used [154], and sound engineers often use a combination of different tactics, or attempt to reduce only one aspect, such as masking of vocals. To date, only Ma proposed an approach that combined multiple methods to address masking issues – rather than to automate a single effect [121]. The study first investigated four metrics for masking in a multitrack mix, and then used a combination of dynamics and frequency processing to perform masking reduction. Subjective evaluation showed that this system produced mixes very close to those of a professional sound engineer in terms of their ability to reduce masking.

8.2.7 Microphone Artifacts

When recording a source with a directional microphone at close range, low frequencies are emphasized. This is known as the 'proximity effect,' and is due to the increasingly low phase gradient compared to the amplitude gradient between the front and back of the membrane. The membrane of omnidirectional microphones, on the other hand, is sealed off in a vacuum on one side. This bass boost could be welcome, or can be perceived as unnatural. However, when the distance between source and microphone varies over time, the resulting change in timbre is almost always unwanted. Clifford and Reiss [339] described a system to detect the proximity effect, using the ratio of spectral flux in low frequencies to spectral flux in high frequencies as an indicator.

Another class of artifacts caused by (mostly directional) microphones is the so-called 'pop' noise that occurs at aspirated plosives in vocals. These are traditionally mitigated using fabric or foam screens between vocalist and microphone, or embedded within the microphone.

However, [340] demonstrated an algorithm to remove these pops from the signal after the fact, using a pair of microphones.

8.3 Humanization

When producing music, it is often necessary to synthesize or sample musical events and sequences. In the DAW, these events can be programmed using a linear sequencer or a piano-roll environment. In these systems, musical meter is preserved using a grid with equal measure. As events are placed onto the grid, they are quantized with user-defined intervals, often down to 1/16th or 1/32nd notes. Perceptually, the quantized events are metronomic and do not necessarily represent the way humans play acoustic instruments. In order to synthesize the perceived naturalness of human playing styles in the DAW, many systems provide a *humanizer* that can be applied to the sequence. This process adds stochastic or aperiodic variables to various aspects of a signal, in an attempt to create a less isochronous output. In percussive sequences these variables primarily control the location of the onset in relation to the computer's click-track, and the amplitude or velocity of each corresponding event.

As workstations typically allocate a global tempo to each session, the most commonly researched aspect of humanization in the DAW has been microtiming, which refers to the small-scale deviations from a beat position in a metrical grid. Formally, this is considered to be the subtraction of an event at time n from a corresponding point on a reference track, as illustrated in Figure 8.2. Here, the reference grid represents a metronome running in parallel with the performed musical sequence. The challenge of the humanization algorithm is to then estimate the distribution at $n + 1$, written as $P(\theta_{n+1})$. Historically, this is applied independently for each of the events in a sequence, based on a distribution centered around the nth grid point, characterized by the parameters μ and σ.

8.3.1 Isolated Sequences

For isolated sequences of events, the distribution from which the onset location is sampled can be generated using a Gaussian window where users are then given control over the extent to which onsets are modulated through a parameter that represents the standard deviation or upper and lower boundaries of the distribution. This means the variability of a musical stream is represented as an Independent, Identically Distributed (IID) Gaussian process. Attempts have been made to increase the naturalness of single-player humanization

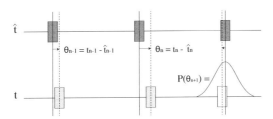

Figure 8.2 Representation of a player, following a metronome. The offset measurements from the metronome are shown as θ_n, where $\hat{t}_n$ is the nth metronomic event and t_n is the nth event as it would be played by a human player.

systems by incorporating some form of intelligent processing into the variate generation procedure.

In an early form of non-Gaussian humanization, fuzzy logic is used to model strike velocity deviation in a humanization system, based on subjective rules derived from domain knowledge [341]. Here, expression is applied to a sequence of MIDI messages using a set of conditions. Similarly, microtiming deviation has been modeled using a number of supervised machine learning techniques. Wright and Berdahl [342] train a number of models on human performance data taken from recordings of Brazilian samba rhythms that have been converted to MIDI. Models such as Locally Weighted Linear Regression, Kernel Ridge Regression and Gaussian Process Regression (GPR) were able to learn patterns in the data with some rule-based pre-training. These techniques are then used to apply the derived microtiming models to quantized sequences, in which they conclude the systems used to model percussive sequences significantly outperform the quantized version, when subjectively evaluated for natural expressivity. Microtiming in rhythmic Brazilian music is the focus of a number of similar studies [343, 344], where models based on the extraction of quarter-notes are developed. Generally, findings show that the characteristics of the expressive timing are closely related to specific metrical positions in the music.

Stables et al. [345] showed that the microtiming in isolated drum patterns can be humanized using a Markov Model. In this case, an event at time n is considered to be conditionally related to its preceding event at time $n-1$. Here various components of a drum kit are modeled concurrently and patterns are considered to sound more naturally expressive when subjectively evaluated against a number of IID and quantized methods.

8.3.2 Multitrack Sequences

In a multitrack project, it is a common requirement for multiple sequences to be programmed for playback simultaneously. Say for example a sound designer requires an ensemble backing track, or a musician needs a sampled rhythm section. In this case, the temporal models suggested in Section 8.3.1 may prove to be problematic due to issues with the phase of concurrent notes. This is because the models that preserve temporal structures of a single stream do not necessarily model the characteristics of group performance when used in a multitrack environment. This can be detrimental to the naturalness of the synthesized signals in the multitrack, due to perceptually unrealistic cues, generated by multiple instances of the algorithm running in parallel.

A solution to this issue was proposed in [346], in which Stables et al. modeled interdependence in a group of musical performers using a Multivariate Markov Chain. Here, the microtiming distribution for player p at time n is conditionally dependent on the microtimings of all of the players in the group at time $n-1$. This allows a series of independent Markov Chains to be given an interdependence factor. In addition to this, the relative dependence between any two players in a multitrack can be manipulated. For example, the bass player could be made to follow the drummer much more closely, or an instrument in a string quartet could be given more expressive freedom. This is made possible via a correction parameter α, which is based on a sensorimotor neuroscience model of group asynchrony [347]. To model this using real-world musical data, the inter-player correction between two streams $\alpha_{i,j}$ can be considered the lagged cross-correlation, in which player i's stream is lagged by a pre-defined number of events n, and correlated with the stream of player j.

8.4 Workflow Automation

8.4.1 Subgrouping

At the early stages of the mixing and editing process of a multitrack mix, the mix engineer will typically group audio tracks into subgroups [108]. An example of this would be grouping guitar tracks with other guitar tracks or vocal tracks with other vocal tracks. Subgrouping can speed up the mix workflow by allowing the mix engineer to manipulate a number of tracks at once, for example by changing the level of all drums with one fader movement, instead of changing the level of each drum track individually. Note that this can also be achieved by a Voltage Controlled Amplifier (VCA) group, where a set of faders are moved in unison by one 'master fader.'

Subgrouping is a technique primarily aimed at enhancing the mixing workflow, by allowing one to change properties such as the level, spectrum, or effect send amount of multiple tracks or sums of tracks. It does not add any new functionality then, but merely saves the engineer from repeating the same operation several times, while retaining a better overview of the session. However, subgrouping also allows for processing that cannot be achieved by manipulation of individual tracks. For instance, when nonlinear processing such as dynamic range compression or harmonic distortion is applied to a subgroup, it will affect the sum of the sources differently than if it were to be applied to every track individually. An example of a typical subgrouping setup can be seen in Figure 8.3.

The subgroup classification problem can been seen as somewhat similar to musical instrument identification, which has been researched extensively. However, in subgrouping classification we are not trying to classify traditional instrument families, but defined groups of instrumentation that would be used for the mixing of audio from a specific genre. For example, in rock music the drum subgroup would consist of hi-hats, kicks, snares etc. while the percussion subgroup may contain tambourines, shakers and bongos. In practice, the genre of the music will dictate the choice of instrumentation, the style in which the instrumentation will be played and the subgroup to which an instrument belongs. It is also worth noting that typical subgroups such as vocals or guitars can be further broken down into smaller subgroups. In the case of vocals, the two smaller subgroups might be lead vocals and background vocals.

Subgrouping Assumptions

Subgrouping is pervasive in music production, yet poorly documented in the literature [108,118,127]. Pestana touched on it briefly in [12,119] when testing the assumption "Gentle bus/mix compression helps blend things better" and finding this to be true, in particular with regard to applying compression to drum subgroups. But this did not give much insight into how subgrouping is generally used. Scott and Kim [55] explored the potential of a hierarchical approach to multitrack mixing using instrument class as a guide to processing techniques, but providing a deeper understanding of subgrouping was not the aim of the paper. Subgrouping was also used by De Man and Reiss in [54], but again only applied to drums and no other instrument types were explored.

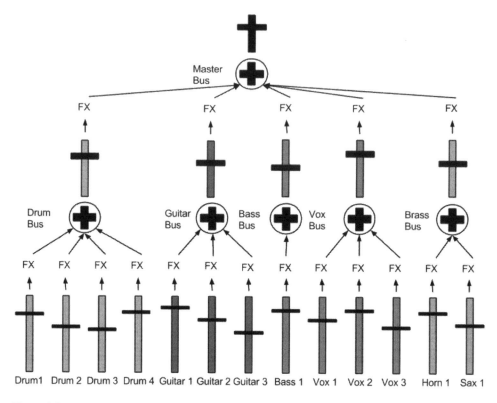

Figure 8.3 Example subgrouping performed in multitrack mixing, from [348]

Most of the research on subgrouping in music production has been performed by Ronan and co-authors. In [348], they looked at 72 mixes of nine songs, by 16 student mix engineers. They analyzed the subgrouping setup for each mix of a particular song and each mix by a particular music production student. They investigated the extent to which the subjects performed subgrouping, what kind of subgroup processing they use, and what effect subgrouping had on the subjective quality of the mix, if any. A complementary approach was taken in [287]. The nature of subgrouping was investigated using a thematic analysis of in-depth, interview-style surveys with ten award winning professional mix engineers. The focus of the survey was on testing hypotheses about subgrouping and identifying subgrouping decisions, such as why a mix engineer creates subgroups, when they subgroup and how many subgroups they use. Together, the two studies provided a wealth of data. Their conclusions, though tentative because of the limited number of participants and a variety of potential biases and confounding factors, may be summarized as follows.

It is very rare to not make use of subgrouping. This may occur when mix engineers are inexperienced, or when they are unfamiliar with the instrumentation or genre. Quantitative analysis of mix engineers' routing habits revealed that tracks are largely grouped based on instrument type, with drums being grouped the most [120, 348]. The number of subgroups correlates strongly with the number of tracks in a session, suggesting that an important reason for subgrouping is to streamline the workflow. Hierarchical subgrouping, where a subgroup contains another subgroup, occurs more frequently with drums and vocals.

Professional engineers confirm this, and further report that the main reasons for subgrouping are to maintain good gain structure, to apply processing to a group of tracks, to create submixes or stems (a stereo mixdown of only a group of tracks), and to be able to quickly monitor or process a particular group of instruments [287]. Interestingly, their survey suggested that professional mix engineers did not generally subgroup to reduce auditory masking. Apart from instrumentation, genre has substantial influence on how they subgroup: it had an impact on the subgrouping practices of six out of the ten professional mix engineers.

As bus compression helps blend sources together, it is the most common effect to be applied [12, 287]. However, no strong correlation is found between the dynamic range compression applied on a subgroup by a mix engineer, and the preference for that engineer's mixes [348]. This goes against the other studies, but it may be because participants in this 2015 study did not have same experience level as engineers interviewed in [119, 287], or because it did not contain enough mixes to see this trend.

Ronan et al. [348] identified a slight subjective preference for mixes which were created using more subgroups, and for mix engineers who tend to create more subgroups on average, especially if processing was applied to the groups. This may mean that the practice of subgrouping is indicative of the experience or performance of the engineer or the time and effort spent on a mix, or that subgroups allow good mixes to be made more easily.

Intelligent Subgrouping

Automation of the subgrouping progress has applications in workflow enhancement – taking care of tedious routing and labeling tasks – and in complete automatic mixing scenarios, for instance when compression needs to be applied to the entire drum bus, or when a group of tracks needs to be considered as a single entity for the purpose of loudness equalization.

Arimoto [183] automatically identifies drum overhead microphones, which typically come in pairs and are often subgrouped hierarchically under the more general drums group [348]. The classification approach makes use of percussiveness, onset count, and coincidence with other (drum) tracks.

Ronan et al. presented an automatic subgrouping algorithm learning from manually assigned instrument class labels [349]. They performed feature selection on 54 multitracks (1467 tracks in total) from the Open Multitrack Testbed [315] to identify 74 audio features to use for automatic subgrouping. The multitracks spanned a variety of musical genres such as pop, rock, blues, indie, reggae, metal and punk. For this, they used a Random Forest classifier, a type of Ensemble Learning method based on growing decision trees [350]. Features that performed below average were eliminated. The multitracks were labeled based on commonly used subgroup instrument types [348]: drums, vocals, guitars, keys, bass, percussion, strings and brass.

For subgrouping, five multitracks were agglomeratively clustered, constructing a hierarchy known as a dendrogram by merging similar clusters. This process is similar to how a human would create subgroups in a multitrack: find two audio tracks that belong together and then keep adding audio tracks until a subgroup is formed. As subgrouping can be likened to forming a tree structure, it makes sense to cluster audio tracks in a tree-like fashion. A maximum of eight clusters were allowed because there were eight labels in the dataset.

If agglomerative clustering is suited to a dataset, linking of audio tracks in the dendrogram should be strongly correlated with distances between audio tracks generated by the distance function. Ronan et al. were able to subgroup the multitracks with a high success rate. Spectrum shape (spectral spread, centroid, brightness and roll off) was important. They listened to incorrectly subgrouped audio tracks and found that these audio tracks suffered from microphone bleed. It should be assumed that background noise, interference or other recording artifacts can affect the accuracy of a classification algorithm.

Jillings and Stables [351] presented another approach to the subgrouping or 'routing' problem using symbolic instrument tags. They combined open linked data and graph theory to infer relationships between labeled channels. Instruments with the same instrument label are grouped by default; other instrument names are assessed on their similarity in terms of overlap between the Wikipedia pages they are linked to, using tags mined from DBpedia,[1] which is a project designed to created linked representations of Wikipedia terms. According to this, the most similar instruments are placed together in one of a specified number of subgroups. The naming of subgroups is also taken care of, by identifying the nearest common subject to each of the instruments in the group.

8.4.2 Processing Chain Recommendation

A large part of the sound engineer's job is configuring parameters on various specialized signal processing devices. Before that, however, the operator or system needs to know which devices need to be used and controlled in the first place. Sure enough, one can predefine a set of effects to be applied to each track based on assumed best practices, as in [54]. But this is a patchy workaround at best, as different situations require a different selection and ordering [108]. In addition, engineers do not seem to universally agree on preferred processor arrangement.

To automate this task, Stasis et al. [128, 352] studied trends in processing chain choices by expert users, when asked to achieve certain timbral qualities. Allowing any cascade using a set of four different audio effects, they found the most popular configurations were (a) just EQ, (b) just reverb, (c) compressor followed by EQ, (d) EQ followed by compressor, (e) just distortion, and (f) EQ followed by reverb [352]. It should be noted that effects like reverberation are often used as a send effect, whereas this experiment only allowed in-line use.

Conversely, a Markov Chain recommender system suggests a list of effects based on one of the descriptive terms that was part of its training set. Objective evaluation of the system confirmed the structure of the space is preserved between the high-dimensional space of timbral adjectives and the lower dimensional space of processing chains. Similarly, the collected data also allows the instrument label to be used to recommend a particular chain. The role of genre proved to be negligible in terms of impact of processing chain.

Analysis of the sequence effects chosen by expert participants further offered interesting insights into mixing practices [128]. As roughly half used just one effect, a further third two effects, and none exceeded a processing chain length of four, it could be concluded short effect sequences are usually sufficient. EQ is also by far the most popular choice for the first effect in the sequence, appearing in this spot twice as often as the compressor.

---

[1] https://wiki.dbpedia.org

8.4.3 Copying Mix Layout

Katayose et al. [29] investigated the possibility of facilitating both processing chain selection and parameter setting, by copying these from a reference mix session. Because of the challenges of the mixing process, and the limitations of putting its desired result into words, they suggested referring to an example is an effective way to convey what the client wants their recording to sound like. Moreover, if this 'template' can simply be pasted, this may save an engineer a significant amount of time, or allow the amateur user to arrive at a much better result.

To test this, a system was devised that applied similar settings to corresponding instruments and sections in a song. Perceptual evaluation then showed that both subjective preference and the ability to correctly guess the 'source' of the mix layout suffered when source and target were very different, for instance in total number of tracks. Clearly, this idea is only practical if a sufficient number and diversity of example mix sessions are available. However, in itself this result is an important finding about mixing in general, showing that preferred settings are highly dependent on the source material they are applied to.

8.5 Live Sound Reinforcement

In this section, we discuss techniques that are primarily deployed in the field of Live Sound Reinforcement. While there is often significant overlap with studio production tools, some aspects are fundamentally different. In contrast to record production scenarios, mixing for live sound is inherently real-time, with a single listening environment but wildly varying listening positions, where the sound engineer needs to account for acoustic summation of several speakers as well as the original acoustic signals.

8.5.1 Feedback Prevention

An audio system is an acoustic feedback loop when the microphones are in the same acoustic environment as the loudspeakers [27], as is the case in a typical live music venue. Whenever the overall gain of this feedback loop goes beyond the nominal level, it generates instability. This phenomenon is commonly known in the audio industry as acoustic feedback, and alternatively named howlback or the Larsen effect [353]. The challenge of feedback prevention lies in maximizing the possible gain that can be applied to useful signals without causing the system to become unstable. This can be achieved by reducing the level of loudspeaker output that is captured by the microphones by pointing them elsewhere (in the case of directional microphones), increasing the distance between loudspeaker-microphone pairs, reducing the microphones in level when they don't capture useful signals, or deemphasizing the frequencies at which the system first becomes unstable. Indeed, the stability of the sound reproduction system needs to be considered at each frequency: a loop gain of 1 (unity) or more at even a single in-phase frequency results in howling at that frequency. Indirectly, reducing the level of noise and reverberation will also allow lower levels and therefore a higher headroom before feedback occurs.

Some of the works on automation of track gain take this constraint into account, effectively maximizing the gain before feedback and allowing automation of the tedious task of gain structure maintenance. In one of their earlier works, Perez Gonzalez and Reiss described a method of automatically calculating the appropriate gain that should be applied in order to maintain a loop gain of one, by mathematically modeling the system [37]. This was implemented and tested in practice with a system consisting of a series of biquad filters, forming a six-band EQ.

Other approaches for mitigating acoustic feedback are known as 'feedback cancellation,' most of which are achieved through static, non-intelligent means and suppress howling after it is detected, rather than preventing it in the first place. These have strong side-effects, such as pitch shifting, distortion artifacts and removing program material, and can be limited in use to speech or systems with a single prominent resonance [37].

8.5.2 Monitor Mixing

An extra complexity of the live sound mixing problem is the presence of personal monitors, which are on-stage loudspeakers intended for the musicians to hear themselves better. Other than the front of house (FOH) speakers, these are customized based on the respective musicians' preferences and needs. Because of their close proximity to the microphones, as well as the usually high sound pressure level (SPL) of the immediate environment, achieving the desired levels of the instruments without feedback occurring is often challenging if not impossible.

Terrell and Reiss focused on this particular situation, treating monitor mixing as an optimization problem with parameters such as the minimum and maximum SPL, gain before feedback, and with each performer's desired levels as the targets [41]. As the targets are unlikely to be reached exactly, a weighting of the preferences determined whose mix is the most important to get (almost) right.

The task of monitor mixing consists not only of getting the balance right, but also the positioning and angling of microphones and monitoring loudspeakers, taking into account the directivity patterns, the distances to performers, or both. For this reason, the position and direction of the monitors as well as instrument amplifiers may also be incorporated as variables to optimize. In [47], Terrell and Sandler took this further by also considering FOH signals and room effects, and their impact on the on-stage mixes and feedback constraints. More generally, target mixes are also defined for positions in the audience.

So far, the focus of these optimization techniques has been on setting the appropriate broadband gains. However, the presented methods can almost trivially be extended to any LTI systems, to include equalization, reverb, and delays [41, 47]. Equalization is particularly important as it can have a profound effect on the maximum gain before feedback, by attenuating those frequencies where the system is least stable. For nonlinear effects such as dynamic range processing, a more complex model is needed. Conversely, the target mixes are defined only in terms of the SPL of the respective sources, meaning features relating to for instance the spectral balance are not considered.

The monitor mixing problem all but disappears when in-ear monitoring is used. In a studio setting, musicians also require monitoring of their own and simultaneously or previously recorded parts. In this case, feedback is rarely a concern as this is typically played back over headphones.

8.6 Reverse Audio Engineering

In this section, we present a number of tools for deconstructing, reconstructing, estimating and repurposing mixing parameters. A number of studies deal with analyzing raw tracks and

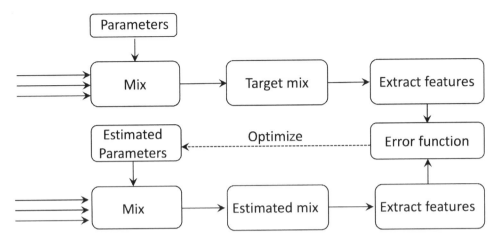

Figure 8.4 Block diagram of the approaches to reverse engineering a multitrack mix

a target mix in order to reconstruct the parameter values of audio effects applied to each track [3, 150, 354, 355]. Their general framework is depicted in Figure 8.4. As shown in the figure, multitrack recordings are mixed using a certain set of parameters, then a feature or a set of features can be extracted from the resulting mix and from a target mix. The distance between those features is computed and is used as an input to an optimization function. This estimates the set of parameters which minimize the distance from the target, so that this particular set can be used to build the approximated mix.

The goal of a study by Kolasinski [3] was to find just constant gains applied to each track, given access to the original tracks and the final mix. It used the Euclidean distance between modified spectral histograms [356] to calculate the distance between a mix and a target sound. A genetic optimization algorithm was then applied to vary gain values until the Euclidean distance converged on a minimum value. The genetic algorithm combined random search with rules inspired by evolutionary processes. However, it came at a huge computational cost and produced less accurate results with an increasing number of tracks. Furthermore, iterative optimization like the genetic algorithm can only converge to an approximate solution of the problem.

Barchiesi and Reiss [354] showed how to reverse engineer both fixed gain and fixed equalization settings applied to each track in a multitrack mix. The technique was then greatly extended in [150]. Here, the settings of time-varying filters applied to each track were derived. Assuming that mixing involves the processing of each input track by a linear time-invariant (LTI) system (slowly time varying processing may be valid if a frame-by-frame approach is applied), then a least-squares approach may be used to find the impulse response applied to each track.

This impulse response represents all the LTI effects that might be applied during the mixing process, including gains, delays, stereo panners and equalization filters. To distinguish these effects, it is assumed that any static level offset is due to fader settings and not a constant gain in the EQ, that equalization introduces the minimum possible delay, that the equalization does not affect the signal level, and that level differences between channels are due to stereo panning. If the multitrack recording was mixed using FIR filters, then the estimation order can be increased until an almost exact solution is obtained. If IIR filters were used, then either one estimates the IIR filter best approximating the finite impulse response, or nonlinear or iterative least-squares approaches are used to derive the filter coefficients directly. Reported results showed the effectiveness of their approach with four, six and eight track multitracks, when both IIR and (up to 128th order) FIR filters were applied, and when there was no knowledge of the characteristics of the (third party, commercial) implementations of the effects that were applied.

Independently of [150, 354], Ramona and Richard [355] also used least-squares optimization to reverse engineer the parameter settings of a mix. However, they focused only on fader levels. They showed that by subtracting the estimate of the summed mix of input tracks from the target track, they can contain an estimate of an unknown additional input, such as an announcer talking over broadcast content in live radio.

One widely used effect for which the least-squares approach is at least partly unsuccessful at reverse engineering the parameter settings is dynamic range compression. The compressor is not LTI, since it is level-dependent, and even the assumption of approximately linear behavior over a short window fails. Barchiesi and Reiss [150] tackled this by assuming that gain envelopes may follow polynomial trajectories within each window. However, though this provided an estimate of the compressor's time varying gain, it still did not give the parameter settings that generated the envelope.

In a study by Gorlow and Reiss [102] this problem was tackled with an original method for inversion of a nonlinear, time-varying system. Here, it is important to note that a compressor is in principle still an invertible operation. That is, unlike clipping or hard limiting, dynamic range compression distorts but does not destroy the information in the original signal. So, the input signal may be expressed as a function of the output signal and the compressor parameters. An iterative optimization procedure is then used to solve this function. This decompressor requires knowledge of the type of compressor that is used, but is fast, effective and applicable to many designs. Later, Gorlow et al. showed that the decompressor can be used to add dynamics regardless of how the audio was originally compressed or even if it was never compressed at all [357], thus providing an alternative to transient modifiers and dynamics enhancers [101, 358, 359].

9

IMP Interfaces

"I automate whatever can be automated to be freer to focus on those aspects of music that can't be automated. The challenge is to figure out which is which."

Laurie Spiegel in Elizabeth Hinkle-Turner, *Women Composers And Music Technology in the United States: Crossing the Line*, Ashgate Publishing, Ltd., 2006.

As users of music production tools, we interact with both hardware and software devices through a range of user interfaces (UIs). Over the years, these interfaces have remained largely consistent and de facto standards have emerged. For instance, almost all mainstream mixing consoles have vertical channel strips, with columns of rotary faders that control parameters such as input sensitivity, equalization and pan, followed by a linear fader which controls the output gain of the channel. Horizontally across each of the channel strips, rows of rotary faders typically allow input channels to be routed to busses, which can be used for additional mixing practices such as subgrouping, or parallel audio effects. These interface blueprints have survived the transition from analog to digital devices, and furthermore have translated almost directly from hardware mixing consoles to digital audio workstations. This trend of reusing legacy interfaces provides familiarity and allows trained audio engineers to easily adapt their workflow when new tools are introduced. However, they do not necessarily provide the most intuition for untrained users, nor do they utilize the capabilities of digital interfaces, or present any perceptual aspects of the audio signal being processed. These interface decisions are very important and have been shown to impact on both critical listening abilities [256] and mixing styles [360]. In this chapter we explore novel methods of interfacing with music production technologies, either through the removal, reduction or

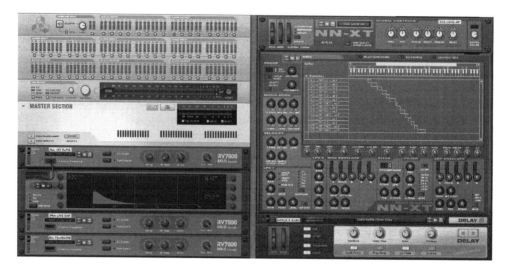

Figure 9.1 UI of Propellorhead's Reaon 9, showing patch chords between virtual rack-mounted synthesizers, sequencers and processing units.

simplification of current interfaces, or through the addition of new mixing parameters and modes of interaction.

9.1 Fundamentals and Concepts

9.1.1 Skeuomorphism

In audio interface design, skeuomorphic elements are particularly common. These are UI components such as controls, labels, backgrounds and widgets that mimic real-world objects. In sound engineering, an overwhelming majority of commercially available software interfaces map control parameters directly from their hardware counterparts. For instance, think of any virtual guitar amp plugin and it is highly likely that you are picturing a brushed metal front-plate, set against a black leather background, with realistic black-plastic potentiometers.

Despite its widespread use, skeuomorphic design has been subject to criticism [361], as it is often perceived to under-utilize digital flat-screen interfaces. Gross et al. [362] point out that critics of this technique tend to consider UI components to be lacking in creativity[1] and for interfaces to often be overwhelming and cluttered.[2] For example, one of the leading rack-mount emulators, Propellerhead's Reason, comes complete with a modular patching interface, which mimics the complex process of connecting synths, sequencers and effects modules together using virtual patch cables (see Figure 9.1).

As a response to this clutter, Sanchez [363] proposes Skeuominimalism, which acts as a trade-off between flat and skeuomorphic UIs. This design philosophy borrows the metaphorical aspects, while preserving important interactions. Sanchez suggests an optimal interface design is "simplified to the point where simplification does not affect usability. And

---

[1] wired.com/2012/01/st_thompson_analog
[2] fastcodesign.com/1669879/can-we-please-move-past-apples-silly-faux-real-uis (Accessed September 2017)

its skeuomorphic affordances are maximized up to the point where it does not affect the simple beauty of minimalism.'"

9.1.2 Automation

By automating tasks in the music production workflow such as mixing and audio effect application (as described in Section 4.7.1), we are applying arguably the most extreme form of interface simplification by effectively removing the parameter space and handing control to an algorithm. For pragmatic reasons, these automated processes are often supplemented with interfaces that allow for user-input. Both [49] and [62] for example, incorporate automatic mixing systems into the DAW via audio workstation plugins, allowing users to apply intelligent processes in-line with a typical workflow. In studies by Perez Gonzalez and Reiss [39, 40], and more recently by Jillings and Stables [120, 268], automatic-mixing and spatialization functions are built directly into a workstation at host-level, shown here in Figure 9.2.

9.1.3 Mapping Strategies

Selecting the appropriate controls for an interface is important, as something as simple as the choice between rotary knobs and sliders can influence the way in which we mix music [364]. Similarly, the interface dictates the ways in which information is visually encoded. For example, channel strips do not necessarily provide an effective way of displaying aspects such as the spatial distribution of a mix across the stereo image [365]. Presenting control parameters in novel ways may involve augmenting the scale of a parameter, providing a new method of visualizing a parameter, or controlling multiple parameters at once.

Parameter mapping refers to a wide range of processes designed to specify the way in which a parameter relates to a sound source. This usually involves manipulating the scale of a variable, such that it more accurately represents some perceptual or acoustic attribute of the source signal. The mapping process can be as simple as using a logarithmic gain fader to control the linear amplitude A_{lin} of a sound signal:

$$y[n] = A_{lin}x[n],$$
$$A_{lin} = 10^{\frac{A_{dB}}{20}}$$
(9.1)

Similarly, mapping can take a single parameter value and convert it to many. Perhaps the most simple example of this is the mapping of a one-dimensional *pan* parameter P to channel gains L and R:

$$L = cos(P\frac{\pi}{2})$$
$$R = sin(P\frac{\pi}{2})$$
$$\text{where } 0 \leq P \leq 1$$
(9.2)

This can be extended to controlling a large number of channels using a low number of controls, for example a bank of channel-gains using a subgroup fader, or nonlinearly controlling a large number of filter parameters using a single slider.

Mapping is particularly important in interface design for virtual instruments. This is because the interfaces for most acoustic instruments are directly connected to the sound source (plucked string, struck key, etc). For this reason, providing natural control of a virtual

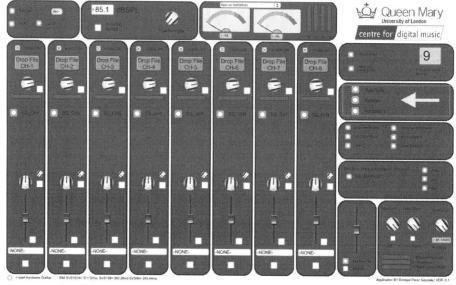

(a) Mixer-Interface developed by Perez Gonzalez et al. [39, 40]

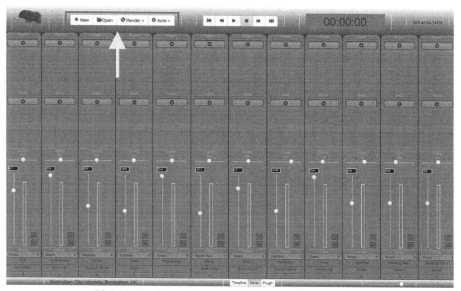

(b) DAW-Interface developed by Jillings et al. [120, 268]

Figure 9.2 Integrated automatic mixing interfaces, with highlighted control regions

instrument is often nontrivial as the relationship between the interface and sound generator needs to be understood by a user. Hunt et al. [366, 367] show that a multiparametric UI with a complex mapping between an oscillator and a physical interface allows users to score highly on a timbral matching task, when compared to one-to-one hardware and software interfaces. The authors make the case for a one-to-many mapping strategy being key in the design of simplified user-friendly interfaces. Arfib et al. [368] present a similar model to account for gesture-to-synthesizer mapping. In this study, the mapping process can be represented as a number of transitions between perceptual and physical spaces.

For the high-level control of audio processing algorithms, mapping strategies are often a key component in processing latent or perceived aspects of sounds. A common method for designing perceptual audio processing tools is to allow effect parameters to control one or more low-level audio features. A number of studies, for example, present methods of controlling the spectral centroid of a sound source, as the feature is highly correlated with perceived brightness [369,370]. This is explored in more detail later in Section 9.3, as a means of using natural language to control music production parameters.

An important aspect of the mapping process is the selection of a control signal, which is used to set or modulate a predefined set of technical parameters in a music production system. These control signals can be categorized based on the way in which they are generated [2].

- *Low-Frequency Oscillators (LFOs)* are time-varying deterministic or stochastic signals, created by a signal generator, which can be used to drive low-level parameters. These are commonly used for modulation in audio effects (phasers, tremolo, auto-wah, etc.).

- *Gestural control signals* are created via some form of transducer, through the conversion of physical movement into a parameter trajectory. These are typically implemented in three-dimensional space, through the use of devices such as video motion capture or inertial movement units.

- *Automation* is gestural control input via a (typically two-dimensional) graphical user interface such as a touch screen device or mouse. This can be done in real time or offline, and is commonly represented as an time-series envelope.

- *Adaptive* control signals, as discussed in Chapter 2, are signals derived from features of a sound source. Features can be derived from the input audio itself (adaptive), or an external source (external-adaptive). Features can be continuous in time such as the f_0 or amplitude envelope, or discrete, such as beats or onsets.

- *Algorithmic* control signals are those which are based on a mathematical process. Historically, this has been particularly common in music composition [1,371,372], and has more recently been applied to the modulation of audio effect parameters [373].

9.1.4 Dimensionality Reduction

Given the relative complexity of music production interfaces, a recent trend in IMP interface design is to reduce the number of technical parameters to a lower-dimensional subset, where the lower number of controls are usually tasked with controlling multiple parameters simultaneously in order to achieve a perceptual transformation. Reduced dimensionality interfaces are typically based on the natural language concepts presented in Section 9.3, by which the reduced dimensionality parameter space is intended to easily apply transformations to the sound, reflective of a description. An example of this is presented by Stasis et al. [374, 375], where the 13 parameters in a parametric equalizer are projected into two-dimensional space in order to control the warmth and brightness of a signal on a flat-screen interface, as shown in Figure 9.3.

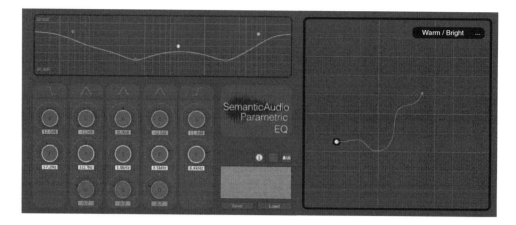

Figure 9.3 An interface with low-dimensional semantic control of EQ parameters, taken from Stasis et al. [375]

Dimensionality reduction exists in a number of forms, all of which aim to preserve some relationship between data points in high-dimensional space, while reducing the complexity of the dataset. Typically, the choice of algorithm depends on the type of data available and the task at hand. In audio research, one of the most common uses of dimensionality reduction is to model musical timbre. In early studies such as that by Grey [376], Multidimensional Scaling (MDS) was used to reduce the number of perceptual dimensions, each representing the extent to which a description can be used to define a sound. The algorithm produces a space in which the data can be easily visualized, while maximally preserving the distances between points in high-dimensional space. This form of dimensionality reduction is commonly used when the dataset consists of pairwise dissimilarity scores. A similar method used extensively in timbral analysis is Principal Component Analysis (PCA), in which eigenvectors of a dataset's covariance matrix are used to apply linear transformations to the axes of the data. This was used by Disley et al. [377] to evaluate the similarity of the timbre of musical instruments. Additionally, Exploratory Factor Analysis (EFA) [378] aims to uncover the latent structure of a dataset by collapsing the variables in a dataset onto a number of factors, based on their intercorrelation. This was used by Zacharakis et al. [379] to identify fundamental descriptive terms in the perception of musical timbre. In this case, both metric and non-metric methods are used in order to identify nonlinear relationships between variables.

In a number of cases, complex relationships between data points cannot be defined using a linear subspace. A number of algorithms have been proposed for these situations, including modifications to PCA such as kernel PCA (kPCA) and probabilistic PCA (pPCA). Here, kPCA is nonlinear manifold mapping technique in which the eigenvectors are computed from a kernel matrix as opposed to the covariance matrix [380], and pPCA is a method that considers PCA to be a latent variable model, making use of the Expectation Maximization (EM) algorithm. This finds the maximum-likelihood estimate of the parameters in an underlying distribution from a given dataset [381].

For situations in which the reduced subspace needs to be organized into classes of variables, such as two groups of data points that represent different types of audio transformation, the supervised Linear Discriminant Analysis (LDA) algorithm maps the data to a linear subspace, while maximizing the separability between points that belong to different groups using the Fisher criterion [382]. As LDA projects the data points onto the dimensions that maximize inter-class variance for C classes, the dimensionality of the subspace is set to $C - 1$. Similarly, embedding techniques such as t-distributed Stochastic Neighbor Embedding (tSNE) [383] perform particularly well for the visualization of groups, as these algorithms attempt to maintain the pairwise distribution of points in high- and low-dimensional spaces by minimizing the KL divergence of both.

Finally, unsupervised neural networks called autoencoders [384] have also been used for the reduction of complex parameter spaces. In this case, a network can learn a reconstruction of its input features with a low-dimensional code-vector set as the central hidden-layer. The network will learn to reduce the data using an encoder, and then reconstruct it using a decoder. This is particularly useful for the reconstruction of parameter spaces due to the codec-like architecture of the network. Stasis et al. [374] show that, in other instances, this reconstruction process has to be approximated using multivariate regression. However, when using a stacked autoencoder, the re-mapping process is embedded into the autoencoder's algorithm, providing a much more accurate reconstruction.

9.2 Abstract Control

Responding to music production interfaces based on skeuomorphism and analog hardware emulations, IMP interface design instead tends towards novel abstract parameter spaces that enable users to achieve creative results using a lower number of actions. These interfaces not only make use of standard graphical input devices, but also explore the use of other modalities such as gestural movement and natural language. In the following sections, we discuss these developments and show how users can process sound using abstract control systems.

9.2.1 Beyond the Channel Strip

The Stage Metaphor

The stage metaphor, first proposed by Gibson [124] as a "virtual mixer" represents the gain and pan parameters of each channel strip as an object in two- or three-dimensional space. In the two-dimensional representation (see Figure 9.4), the horizontal axis represents pan, and the vertical axis represents gain. In the three-dimensional representation, the additional axis represents a number of filter parameters such as cutoff frequency and gain. This form of visualization is intended to represent a more intuitive mixing environment based on its likeness to a sound stage (i.e., musicians standing at different positions on a virtual stage). If we consider the gain (g_{ch}) and pan (p_{ch}) parameters in a channel strip environment to be orthogonal, then calculating the gain of an object on a two-dimensional stage is mathematically equivalent to converting between the Cartesian and Polar coordinate systems:

$$
\begin{aligned}
g_{st} &= \sqrt{g_{ch}^2 + p_{ch}^2} \\
p_{st} &= tan^{-1}\left(\frac{p_{ch}}{g_{ch}}\right)
\end{aligned}
\tag{9.3}
$$

$$
\begin{aligned}
g_{ch} &= g_{st} \cdot cos(p_{st}) \\
p_{ch} &= g_{st} \cdot sin(p_{st})
\end{aligned}
\tag{9.4}
$$

Here, the gain in the stage visualization (g_{st}) is the distance from a listening point, i.e., the magnitude of the two-dimensional gain/pan vector, and the position in the stereo image (p_{st})

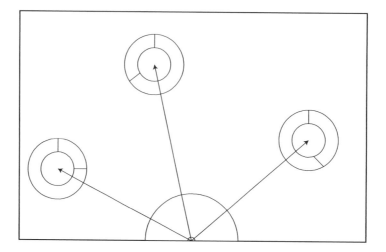

Figure 9.4 Two-dimensional representations of the stage metaphor for music mixing

is the rotation or phase of the object, as observed from the listening position. This conversion between stage and channel strip methods assumes a linear gain and panning law.

This method of mixing has been shown to provide very similar results to the channel strip method, when subjects were given the task of matching the gain balance and pan positions of a reference mix [385]. Participants are able to converge on the same relative mix with no significant variation in speed or accuracy. Additionally, similar studies show the stage model outperforms the channel strip when performing complex mixing tasks [386], particularly when trying to match the panning positions of individual channels in a mix. The sound stage representation has been implemented in a number of alternative stand-alone mixing tools [387, 388]. Rodriguez and Herrera [389] extend the model to work as a VST plugin in a traditional DAW, based on Gibson's original three-dimensional representation. Here, loudness, stereo panning and frequency are mapped to parameters of the spherical track objects.

A reported issue with the stage visualization is the potentially cluttered interface, particularly when objects occupy similar gain levels and pan positions [390]. As more objects (tracks) are added to the mix, the stage representation becomes harder to navigate due to overlapping boundaries. To address this, Gelineck et al. [391] explored new methods of presenting channel information using features of the circular objects. Channel activity, level and auditory brightness were mapped to features such as the length of a circular line around the object, the roughness of the object's boundary and the visual brightness of the object. While these features can provide additional information, the authors suggested that too much visual information can be overburdening to the user, adversely affecting the usability of the system.

Mycroft et al. [392] trialed an eight channel prototype mixer that used a novel approach to the mixer design to address the cognitive load of channel strip designs. The mixer used an overview of the visual interface and employed multivariate data objects for channel parameters, which can be filtered by the user. They showed improved results in terms of visual search, critical listening and mixing workflow. They attributed this to the reduction in

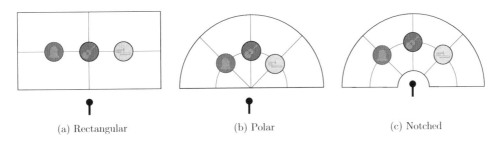

(a) Rectangular   (b) Polar   (c) Notched

Figure 9.5 Stage view representations explored by De Man et al. [394]

both the complexity of visual search and the amount of visual feedback on the screen at any one time,

Dewey and Wakefield [393] investigated methods to prevent this reported clutter, by exploring a range of methods for visualizing the track objects in the mix. They considered six visualization styles: text only, black and white circles, individually colored circles, circles colored by group, instrument icons, and circles with dynamic brightness. The study showed that track selection times increased across the board when the number of channels increases, however the text only objects provided the fastest reaction times from participants in the study. Interestingly, while the simplistic interfaces provided the most efficient method for mixing, users actually expressed subjective preference for the more abstract interfaces. Recently, this philosophy was also applied to a virtual reality implementation of the stage view interface, where it was found the UI allowed for equally efficient mixing practices [390].

In a study by De Man et al. [394] the stage metaphor interface was explored further by testing the performance of three stage representations: (a) a rectangular design that used Cartesian coordinates, (b) a semi-circle design that used polar coordinates, and (c) a notched semi-circle design, which used polar coordinates. The study showed that users preferred the notched view as the polar coordinates are more representative of human perception, and the notch allows for more control over the pan position when a source has a high gain setting. The three designs are presented in Figure 9.5.

Spatialization

The stage metaphor has been extremely successful in providing interfaces for surround sound and spatialization systems. This is particularly salient given that the spatial distribution is often poorly represented in a standard mixing environment. Pachet and Delerue [28] proposed an early method for spatialization of sound sources in a mix. Their system MusicSpace allows the user to move a virtual avatar 'on the fly,' meaning that users can manipulate the audio during playback. The model is based on a constraint language and solver, and uses a stage-like interface in which the listener avatar can be moved in any orientation from the instrument objects, including centrally. Similarly, AWOL [395] is a system that utilizes features of the stage metaphor to represent the sounds in a 3D image. In this system, the sound propagation of each source is presented through concentric circles centered around each channel object, a visual aid that is not possible in a standard environment. The system is designed to run alongside a standard DAW, allowing the user to visualize the direct relationship between stage and channel strip interfaces.

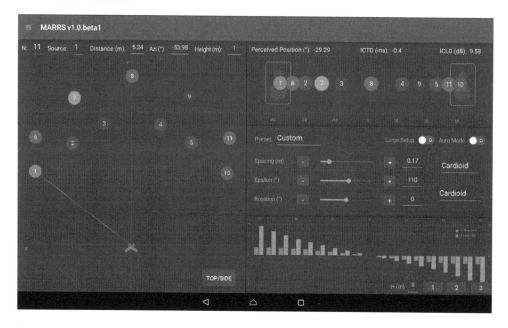

Figure 9.6 The MARRS GUI for microphone array design, proposed by Lee et al. [396]

The stage model has also been used successfully for the design of microphone arrays. This aids the process of achieving a desired spatial distribution of sound sources in a stereo image, by informing the user where to position multiple microphones during recording. The Microphone Array Recording and Reproduction Simulator (MARRS), developed by Lee et al. [396], presents users with an interface which guides them through this process through manipulation of objects on a stage (see Figure 9.6). The system uses psychoacoustic principles such as adaptively applied linear image shift factors (originally presented by Lee and Rumsey [397]), in order to provide an optimal position and subtend angle for a near-coincident stereo pair of microphones.

Multi-touch Surfaces

Due to the ubiquity of touch-screen interfaces, almost all of the high-profile pro-audio manufacturers have released software applications which interface with one or more of their hardware mixing consoles via a mobile or tablet device. This can increase efficiency during tasks such as live sound engineering, as the engineer is able to control channel strips while walking around the venue, or while discussing requirements with musicians on stage. However, most of these software packages still offer a one-to-one mapping between the touch surface and the channel strips or effects racks.

An alternative to this is the recent work of Gelineck et al. [388, 391, 398], which extends the stage metaphor for multi-touch surfaces, mobiles and tablets. The authors point out that a touch screen interface is a particularly impactful use of the stage metaphor as directly touching the interface components gives the user a sense of controlling the sound sources directly. These studies show that a number of adjustments to the stage metaphor are necessary to adapt the model for a small, and often cluttered screen. Firstly, the objects are given a

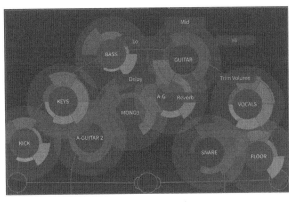

(a) Gelineck et al. [388]

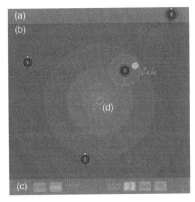

(b) Carrascal et al. [399]

Figure 9.7 Two multitouch interfaces that make use of the stage metaphor for music mixing

pie-chart style interface, in which the circular widgets are divided into equal segments, each of which controls an assignable audio effects parameter, as shown in Figure 9.7. To achieve this, a user can drag from the center of the object in a chosen direction and the parameter increases. Similarly, the authors implement a layering system to avoid clutter. This means that subgroups can be shown or hidden to avoid congestion. A method for making the metaphor more acoustically realistic is implemented by reducing the gain, removing high frequencies, and increasing reverberation as the object moves further away from the listener position.

Carrascal and Jordà [399] presented a similar method for controlling a mix using a multi-touch table interface. The key difference here is the use of a stage environment for each of the auxiliary sends in the mix. Unlike other interface designs, the listening position is centered on the UI panel, suggesting objects can be placed behind a listener (see Figure 9.7).

The stage metaphor has also been applied to smart tangibles, which are small hardware devices, often known as blocks, which contain a range of sensors and input devices. These are often combined with multi-touch surfaces to make immersive tactile interfaces such as the ReacTable [400]. In music production, these blocks can be manipulated and moved in some predefined space in order to achieve a mixing task. Gelineck [401] proposes a number of methods for assigning the parameters of a series of smart tangibles to parameters in a stage metaphor, which all complement the placement of a physical block within a semi-circle representing the stage. Additional components in the channel can be controlled by rotating and pressing the block to control filter, reverb, compression and delay parameters. Similarly, the Cuebert mixing board [402] is a reconceptualization of a mixing console, using a combination of multitouch surfaces and physical analog controls such as rotary knobs, faders and a keypad. The board was designed specifically for musical theater use, and has a highly adaptive interface based on cues (points in a production when the environment needs to change based on an audio event). The authors conducted a thorough qualitative survey using semi-structure interviews, and identified a number of key requirements such as

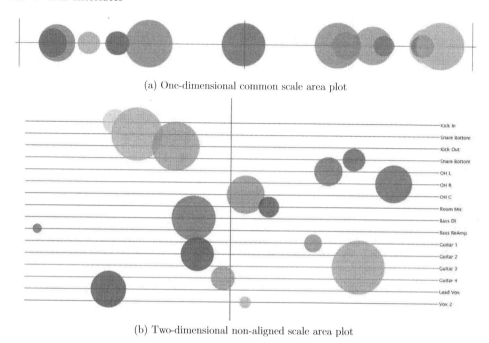

(a) One-dimensional common scale area plot

(b) Two-dimensional non-aligned scale area plot

Figure 9.8 Two interface examples presented by Dewey et al.

a preference for one-to-one mappings, a dislike of layers (hierarchical subgroups), and the need for a central panel of frequently-used parameters.

Several studies noted that multitouch surfaces quickly become cluttered, as objects often obstruct each other. Some parameters, such as effect parameters, are not as easy to control when using smart tangibles. Finally, experienced users still often prefer to see the channel strip interfaces in order to properly understand the mapping process.

Alternative Channel Strip Representations

Dewey and Wakefield [403] presented a number of alternatives to the stage metaphor, based on data visualization first principles. These typically include a projection of channel strips onto a two-dimensional space, in which the position along the x-axis is used to represent the position of the corresponding sound source in the stereo image. A number of derivatives were proposed, including circular objects, stems and icons, whereby the size or elevation of the object represents the track gain. These single-line models were compared to a staggered approach, in which each channel has its own fixed position along the vertical axis (see Figure 9.8). The authors show that the model with the highest performance when applied to an eight channel matching experiment was a non-vertically-aligned scale (Figure 9.8b) which used icons to represent the channels in a mix.

Gohlke et al. [404] presented a range of alternative visualizations to the standard waveform view in the DAW GUI, including the use of music notation and abstract spectral representations. The authors argued that there is an overwhelming amount of unnecessary visual information in the standard DAW interface and that more efficient methods for conveying information are available while using less on-screen real estate. One of the focuses

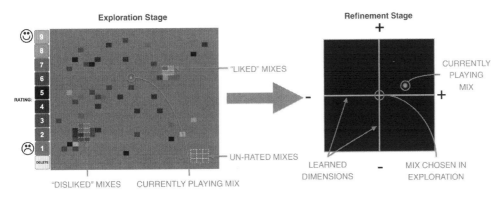

Figure 9.9 The Mixploration system for intelligent mixing, proposed by Cartwright et al. [405]

of the study was the reduction of the amount of visual real-estate occupied by waveform views. The authors showed that in order to prevent this, waveforms can be moved closer together with variable opacity, or can be reduced and navigated using a 'lens' view, to magnify areas of a region.

Similarly, Mycroft et al. [386] suggested that the overwhelming amount of visual information in a DAW can be mitigated via the use of Dynamic Query filtering. This technique allows a user to move a slider that dynamically adjusts the visual components of the UI. This allows for rapid exploration of the space, and the authors showed it increased productivity. In this particular case, a range of clickable objects were used to filter out channels with certain pan positions or gains.

Mixploration [405] (shown in Figure 9.9) provides a novel interface for intelligent channel mixing and equalization. The system separates the mixing process into two stages: exploration and refinement. The exploration stage allows the user to navigate a range of predefined mixes, at which point the user can rate them based on how relevant to a desired target mix they are. The refinement stage then allows the user to fine-tune the selected mix using an abstracted 2D space, affecting equalization and fader balance. The result is a mix derived through similarity, as opposed to the traditional gain balancing process.

9.2.2 Digital Audio Effects

Digital audio effects are used in in conjunction with fader balancing, panning and routing to form the basis of mixing. Effects are generally arranged in series on a channel strip to form a processing chain, or used on busses for parallel processing.

Equalization is a particularly prevalent audio effect, due to its wide range of applications in an audio engineer's workflow. Users tend to interact with these systems through knobs and sliders that are connected to center frequency, gain and bandwidth parameters of filters, often with movable UI widgets overlaid on a frequency analyzer. A number of authors [256, 406] have considered the relative complexity of the standard equalizer interface, and suggested that simplified designs may encourage users to focus on the audible properties of a signal, as opposed to the visual information provided by a frequency analyzer. Dewey et al. [406] evaluate various novel interfaces, which incorporate spectral information in addition to, or as a replacement for spectral curves.

Loviscach et al. [407] extend this traditional idea for a five-band parametric model, by allowing the user to use freehand drawing to achieve the desired magnitude response. To do this, evolutionary computing is used to match the curve to a relevant parameter setting. The study shows that this can reduce time taken for configuration compared to the traditional methods. Heise et al. [408] develop this use of evolutionary computing further, using a DAW plugin which learns from user settings. The models are tested with hearing impaired users, allowing them to configure their own settings, without the requirement for music production experience.

An alternative approach is to reduce the dimensionality of the parameter space, so that it can be controlled using a two- or three-dimensional UI component. Both Mecklenburg et al. [17] (SubjEQt) and Sabin et al. [263] (2DEQ) employ this method in order to create two-dimensional equalizer interfaces. The authors map a 40-band graphic EQ into a two-dimensional space, by gathering user settings and using PCA to present them in two dimensions. 2DEQ additionally makes use of a self-organizing map to organize the spatial configuration of settings in the low-dimensional interface.

As mentioned in Section 9.1.4, Stasis et al. [374, 375] make use of a stacked autoencoder to map between high- and low-dimensional equalizer parameter spaces. Using this neural-network-based method allows for effective mapping between the two spaces as the model includes an encoder and decoder. The study compares the technique against a number of other dimensionality reduction (PCA, MDS, etc.) and reconstruction (linear regression, kernel regression, etc.) methods, to show that the model outperforms the others in terms of subjective quality. These low dimensional interfaces are commonly used in conjunction with descriptive terminology, in order to assign perceptual labels to the abstract dimensions. Models that make use of these labels are discussed in detail later in the chapter (see Section 9.3.5), as they are aligned with the use of natural language in sound production. This model has also made its way into the mainstream audio software industry, with systems like the Waves OneKnob[3] series projecting a number of audio effect parameters onto a single dimension, and controlled using a subjective term.

9.2.3 Metering

Intelligent metering tools have the potential to make mixing easier for novices and to improve the efficiency of experts. Traditional DAWs use metering tools to visually present important information about a mix. This often includes the representation of a signal's level on a linear meter, or the spectral content of a signal using a frequency analyzer in an equalizer. Intelligent metering tools attempt to present high-level perceptual or musical features of signals, in order to more accurately represent the relevant aspects of a mix. This concept was demonstrated by Jillings and Stables [120], in which regions in a sequencer view of an intelligent DAW have assignable waveform representations, including low-level audio features and some primitive measures of loudness.

An example of this is the presentation of perceived loudness, as opposed to the signal's intensity. This involves the approximation of various components of the human auditory system, including models of the outer and middle ear, computed excitation patterns, and spectral/temporal integration. Historically, this has been a difficult measurement to present

---

[3] www.waves.com/bundles/oneknob-series

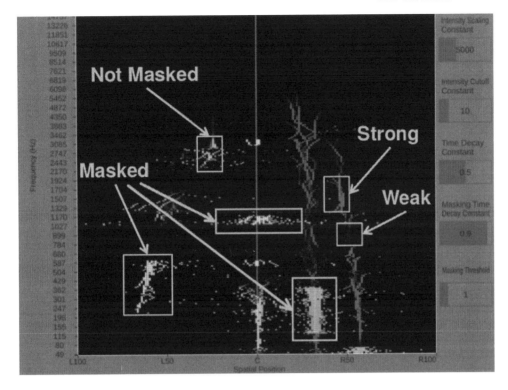

Figure 9.10 The MixViz system for stereo masking visualization, created by Ford et al. [410]

in a DAW due to the computational complexity of various components in the model. Ward et al. [146] presented an experiment to evaluate the extent to which excitation-based loudness models can run in real time, culminating in a DAW plugin that allows users to visualize binaural loudness. The plugin considers the loudness on a single channel, and is able to run in real time due to a number of optimizations applied to both the Glasberg and Moore [270] and the Chen and Hu [409] loudness model.

A key part in our understanding of auditory loudness is spectral masking, which occurs when the interaction of sources in a mix causing the perceived loss of spectral energy in one of the sources (see Section 8.2.6 for a more detailed definition). MixViz, a system developed by Ford et al. [410], addresses this issue by presenting visual information about inter-channel masking to users in real time, allowing the engineer to reduce the loss of salient spectral content during the mixing process. The system extends the Glasberg and Moore model of loudness [270] to account for spatial interaction between sources. This is visualized through a two-dimensional interface (shown in Figure 9.10), where stereo position is presented along the horizontal axis, and frequency along the vertical axis. The system currently doesn't provide direct manipulation, but users can use the visual information to apply corrective panning and equalization.

A number of systems focus on the presentation of salient frequencies from a given source [92, 93, 96], in order to provide the mix engineer with a starting point for equalization, or to inform intelligent processes such as adaptive effects, or automatic mixing algorithms. These systems do not necessarily propose graphical interfaces, but they can be used as a starting point for the application of effects. Salient frequencies are those that should be addressed by a user for the purpose of spectral balancing, due their prominence in a signal. Bitzer and LeBeouf [92, 96] identify these frequencies by comparing a long term average spectrum with frequencies selected by experienced users. They then proposed a number of methods such as identifying the roots of a linear predictive coding (LPC) algorithm to uncover resonances in the spectral envelope, or by applying peak-picking heuristics to a Welch Periodogram. Dewey and Wakefield [93] extended this idea by eliciting user ratings from an equalization experiment.

Some approaches in this area derive high-level qualities from low-level audio features. De Man et al. [82] presented methods for the analysis of total reverb amount in a multitrack mix by analyzing the loudness and decay time of an 'equivalent impulse response.' To test this approach, the authors provided a number of mixes with variable reverberant properties, produced by experienced engineers, and elicit feedback from participants in a multiple stimulus listening test. The study confirmed that the settings of reverb effects has a large impact on the quality of a mix, and that perceived reverb amount can be estimated based on these objective measures. It also suggested a likely range of suitable values could be identified for a certain effect parameter.

In a number of studies, Fenton et al. presented methods for estimating *punch*, i.e., the dynamic clarity of transients in a signal [135, 411, 412]. The process comprises transient extraction, loudness pre-filtering, band splitting, onset detection, and energy weighting, resulting in a score which correlates well with subjective tests.

9.2.4 Gestural Control

Naturally, we consciously and unconsciously move our body to musical stimuli, whether it is tapping our feet to the beat, or moving our arms with the movement of a strings section [413, 414]. For this reason, it seems sensible that gestural control over music production parameters would form a natural mode of interaction. Gestures can be categorized using a number of canonical types [415, 416]: *static* gestures which require no significant bodily movement of such as taps; *dynamic* gestures which require instantaneous bodily movement; and *spatiotemporal* gestures, which rely on multiple measurements over time. These gesture types can be broken down further into subcategories based on their underlying representation [417]. If a gesture is intended to emulate an existing action such as playing an instrument, it is considered *mimetic*. If a gesture is selected from predetermined set of existing gestures, where each has a specific function, it is considered *semaphoric*. If a gesture is specifically designed to allocate objects such as pointing, then it is considered to be *deictic*. Each of these have implications on the use of gestures for given situations and how intuitive they are for an end user.

Balin and Loviscach [418] pointed out that music production interfaces have a surprisingly small number of gestural controls, as interfaces are generally graphical and based on components that do not move linearly (such as rotary faders). They conducted a survey in which participants were able to control DAW parameters with two-dimensional single- and multi-point gestures. The study showed that subjects chose from a wide range of tap-based, swipe-based and symbiolic gestures, when they were asked to assign gestures to a range of common DAW controls such as *split selected region, jump to end,* and *undo last action* (see Figure 9.11).

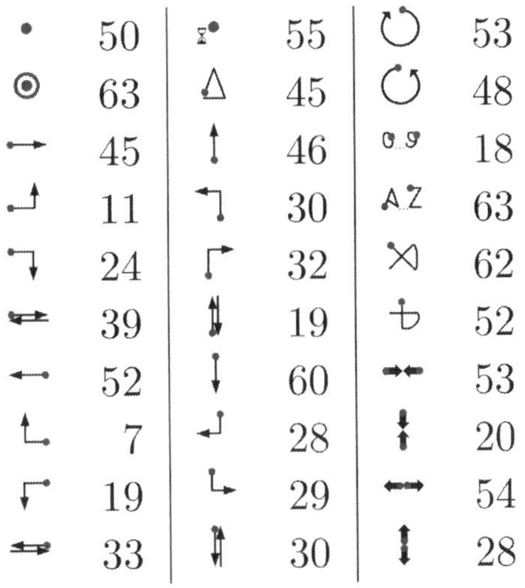

•	50	꜒•	55	↻	53
◉	63	△	45	↻	48
→	45	↑	46	∿	18
⤷	11	↰	30	A.Z	63
↳	24	⌐	32	⋈	62
⇄	39	↕	19	♭	52
←	52	↓	60	→←	53
⌐	7	↵	28	↕	20
⌐	19	⌐	29	←→	54
⇌	33	↕	30	↕	28

Figure 9.11 Gestural movements frequently chosen by participants in [418]

But implementing gestural controls for music production can be relatively arduous for a user of the system, as they are required to quickly learn a number of complex gestural-to-auditory mappings, while also focusing on the sound of the mix. Marshall et al. [419] cited a number of key design factors in developing gestural interfaces in the context of sound spatialization:

(a) *discrete or continuous* controls should be confluent with the parameter they are manipulating;
(b) the *resolution* of a control is important given the variability in ranges of audio processing parameters;
(c) the *control rate* of a signal (i.e., the rate at which it changes) should be related to the control being used to modulate it;
(d) *separability and integrality* should be considered when designing gestural systems, i.e., if parameters are closely related, then similar or related gestures should be used to control them.

Verfaille et al. [2] proposed a number of mapping strategies for gestural control of audio effects. The authors focused specifically on the use of gestures with adaptive effects, whereby the effects' parameters can be controlled using features of the input audio signal. A number of modular topologies were proposed based on the use of gestures in sound synthesis (see Figure 9.12). Here, gestures and audio features were considered to be inputs to a system, which need to be processed and combined in order to map them to the parameters of an audio effect. The most basic system (Figure 9.12) extracts features from both gestures and input sounds in a single module, then scales them to be used as control values for the effect. To provide more intuitive control, allowing the user to more clearly understand the relationship between the gestures and the output of the effect, gestural and audio feature extraction can be separated into independent blocks, and used to control

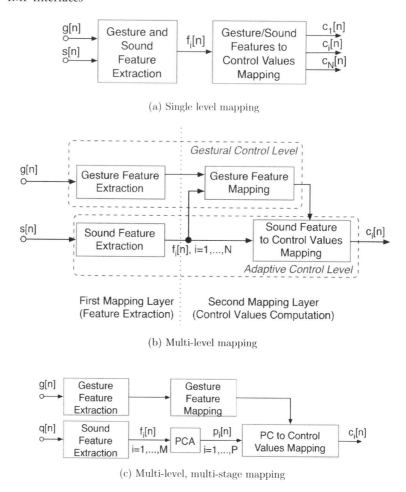

(a) Single level mapping

(b) Multi-level mapping

(c) Multi-level, multi-stage mapping

Figure 9.12 Mapping strategies for gestural control of adaptive audio effects proposed in [2]

different parameters of the audio effect. A complex, but potentially more powerful method for gesturally controlling adaptive audio effects is to allow the gestural and audio control signals to modify each other (Figure 9.12). This method of automation provides much more flexibility, however it may be less intuitive. To further augment these mapping processes, the authors described a number of additions to the mapping topologies, such as dimensionality reduction and multiple control layers.

For gestural stereo mixing, Lech and Kostek [420] presented a system for mapping free-field hand gestures to mix parameters, which can be used with or without visual feedback. This works by analyzing a video feed and tracking the shadow of a users hand using a standard webcam. Gestures are tracked using fuzzy logic for dynamic movements, and a support vector machine (SVM) for static positions. The study showed mixes using the gestural system are comparable to mixes using a traditional interface, with some apparent trends, such as more extreme usages of filters.

Ratcliffe [365, 421] also presented a method for hand gesture motion-controlled spatial mixing. The system combines a Leap Motion device, which allows the user to move a sound source laterally and horizontally, while using TouchOSC to select channels. LAMI [422] also used a Leap Motion to control objects using the stage metaphor. The device tracks rotational and muscular movements, and twisting the arm controls parameters such as rotary faders. The authors commented that this system is still in its infancy and while provides entertainment, does not perform as well as an on-screen DAW for music production.

An alternative device for gestural audio mixing is the Nintendo Wii controller. This is a hand-held device with a number of inertial movement units and buttons built-in, including x, y, and z-axis accelerometers, and an IR sensor. Selfridge and Reiss [423] used this device to manipulate a range of mixing parameters including gains, stereo position and audio effects using one-to-one mappings, but showed that these controllers have many limitations when applied to music mixing. Morrell et al. supplemented Wii controller-based mixing tools with auditory feedback systems in order to make them more usable to a general audience [424]. This involved providing information on sound source selection in the form of spoken descriptions of the selected sound, and by changing the perceived loudness of the selected source against other sources in the mix.

Fuhrmann et al. [425] showed how multiple Microsoft Kinect sensors can be used to allow synthesis parameters to be adjusted using gestures in real time. The authors suggested the method is low latency, low cost, intuitive and precise, and allows the user to control the system without issues that arise based on occlusion (e.g., an object or body part blocking a gesture during performance). The study presented a number of methods for extracting useful data from human movement, with an emphasis on methods for merging data from multiple sensors. To make the movement data more accurately control synthesis parameters, a number of suggested one-to-one nonlinear mappings were applied to the control data. For example, linear axes are scaled and converted to logarithmic, exponential or sinusoidal scales to manipulate the f_0 and timbre trajectories of a sound source.

Modler [426] extended this idea by incorporating machine learning to identify hand-gestures, before mapping them to sound transformations. The users of the system interact with synthesis parameters using a sensor-glove, which outputs hand and finger movement data to a neural network trained on annotated hand-motion data. The system more accurately classifies gestures, which are then translated to synthesis parameters during the mapping procedure. Naef and Collicott [427] also used a glove as an input sensor, with an emphasis on music performance in virtual reality environment. Here, the glove does not contain sensors, but is tracked by a depth-camera and mapped to DAW controls. Rotations control gain parameters for example and can be mapped to audio effects. For additional feedback, haptic devices are placed on each of the finger-tips, so the user is provided with a tactile representation of the mixing space.

A similar device that focuses on the spatial positioning of sound sources is MyoSpat, presented by Di Donato et al. [428]. This system utilizes the functionality of the Myo gestural control armband, which allows both lateral movement and hand gestures. It has predominantly been used for the manipulation of sounds, and their position in a stereo image during live electronic music performance.

Similarly, Drossos et al. [429] presented a hand-based gestural mixing interface in which custom boards comprising accelerometers, push buttons and IR proximity sensors are connected to an Arduino microcontroller. The device transmits MIDI data, which in turn is able to manipulate gains, pan positions and audio effect parameters of a DAW session.

Spatial positioning is controlled using movements along a two-dimensional plane, and effect parameters are controlled using hand rotations. The authors reported that a sensible range for the proximity sensor was 15–60 cm, which in turn is mapped to 0–128 linearly to cope with the 8-bit range of a MIDI device.

Wilson et al. focused on gestural control of EQ and compression respectively [416, 430] . Here, gestures are mapped to semantically defined presets, which are determined by the SAFE dataset [184]. Gestures are initially elicited through a user-survey, and then implemented via a two-dimensional GUI.

9.2.5 Haptics and Auditory Displays

Accessibility in music production is often overlooked due to the relative complexity of the platform and the modularity of most interfaces. This issue has been of interest to pro-audio companies such as Avid,[4] who provided a historical overview of the compatibility issues with native Apple voice over utilities. The company suggested that currently, the biggest problem for audio description tools is caused by badly labelled plugin parameters. In academic research, a number of studies were concerned with the development of music production systems that accommodate visually impaired sound engineers. These typically make use of auditory and haptic feedback as a means of informing and allowing an engineer to navigate the mix space.

An early example of this is The Moose [431], a hardware device very similar to a standard computer mouse but with haptic feedback allowing visually impaired sound engineers to interact with an audio workstation. In this case, mouse-over actions on faders and buttons would generate vibrations, allowing the engineer to physically feel the on-screen GUI parameters. The idea of a embedding haptic feedback into a computer mouse was extended by Geronazzo et al. [432], whereby spatial audio cues are used concurrently to improve a user's ability to recognize objects when their visual modality is obstructed, and Hlatky et al. [433], in which pseudo-haptic feedback was provided for fine-tuning virtual faders. A similar system for the mapping of auditory data to the tactile modality is Haptic Wave [434], a hardware tool designed to replace the visual aspects of audio editing by representing the dynamic temporal information as kinesthetic information in the haptic domain. The system allows users to locate a position the mix by scrubbing a physical slider, which provides stronger tactile feedback when the waveform has a higher relative amplitude, shown here in Figure 9.13. A similar method was presented by Karp and Pardo [435], in which a chain is placed onto a magnetic surface (shown in Figure 9.13). In contrast to Haptic Wave, the user can actually interact with the equalization curve by modifying the shape of the chain. The model is implemented as a DAW plugin, and uses a series of low-cost components. For instance, the chain's movements are tracked using a standard USB camera and OpenCV [436].

Metatla et al. [437] highlighted a number of significant limitations of standard DAW UIs for visually impaired engineers. One of the main barriers is the overwhelming reliance on graphical components such as meters, waveform views and graphs, none of which can be used in conjunction with a screen reader. The authors therefore addressed two key problems using sonification: constructing an automation curve, and reading the clipping information

---

[4] www.avidblogs.com/music-daw-software-for-blind-and-visually-impaired-audio-professionals

(a) Haptic Wave [434]

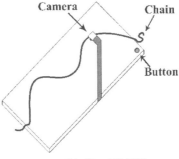

(b) HaptEQ [435]

Figure 9.13 Two methods for the provision of haptic feedback during equalization

from a peak level meter. For automation, a number of sonification techniques were suggested including mapping the y-axis to the pitch of a pure tone, and creating 2 tones, where one is used as a reference. To identify clipping in an audio track, the clipped regions of a waveform were used to apply frequency modulation to a pure tone.

9.3 Natural Language

Sound engineers typically use descriptive language to refer to the way a component in a mix (including the mix itself) is perceived [184, 438]. While the vocabulary is typically very diverse, a subset of terms are very common, and there is some degree of consensus on their meaning. Still, it is a subjective phenomenon, which makes it nontrivial to develop systems that can understand and replicate the sounds that the engineer is describing. This problem is reflective of a wider issue in affective computing, which is that computers and humans behave in different ways when presented with data. Human engineers are able to, consciously and unconsciously, position sounds in an emotive conceptual framework, biased by a history of sound perception. Additional factors such as the genre and mood of the music, other sources in the track, and the environment in which the engineer is listening can all influence the perception of the sound. Computers on the other hand, at least given current technology, react only to very specific instructions, and do not understand the context in which the sound is being presented. For this reason, computational models of timbral adjectives have been of interest to research communities for the last few decades. If we are able to develop effective methods for timbral control, not only will we know more about the ways in which the brain perceives audio signals, but we will then also be able to develop abstract, intuitive methods of manipulating them and computationally model production decisions.

9.3.1 A Shared Vocabulary

In audio engineering, there is a corpus of terms which are widely accepted as having correlated spectral properties. We often refer to this as a vocabulary of descriptors. Most experienced audio engineers and musicians tend to be in agreement regarding the perception of these descriptors, and because of this, these terms have formed the basis of audio effects presets across a wide range of software tools. However, it can be the source of

Table 9.1 An excerpt from the table of spectral descriptors and their associated frequency ranges, according to various sound engineering handbooks, from De Man [114]. The complete table is listed in Appendix A

Term		Range	Reference
air		5–8 kHz	[127, p. 119]
		10–20 kHz	[439, p. 99]
		10–20 kHz	[108, p. 211]
		11–22.5 kHz	[127, p. 26]
		12–15 kHz	[440, p. 103]
		12–16 kHz	[115, p. 43]
		12–20 kHz	[439, p. 25]
		12–20 kHz	[441, p. 108]
		12–20 kHz	[442, p. 86]
anemic	lack of	20–110 Hz	[108, p. 211]
	lack of	40–200 Hz	[127, p. 119]
articulate		800–5000 Hz	[127, p. 119]
ballsy		40–200 Hz	[127, p. 119]
barrelly		200–800 Hz	[127, p. 119]
bathroomy		800–5000 Hz	[127, p. 119]
beefy		40–200 Hz	[127, p. 119]
clarity		2.5–4 kHz	[442, p. 86]
		2.5–5 kHz	[285, p. 484]
		3–12 kHz	[108, p. 211]
		4–16 kHz	[127, p. 26]
fat		50–250 Hz	[108, p. 211]
		60–250 Hz	[127, p. 25]
		62–125 Hz	[115, p. 43]
		200–800 Hz	[127, p. 119]
		240 Hz	[285, p. 484]
presence		800–12000 Hz	[127, p. 119]
		1.5–6 kHz	[439, p. 24]
		2–8 kHz	[115, p. 43]
		2–11 kHz	[108, p. 211]
		2.5–5 kHz	[285, p. 484]
		4–6 kHz	[127, p. 25]

misunderstanding due to differing definitions [6, 117]. Table 9.1 shows a selection of a list presented by De Man [114], included in full in the Appendix. It provides a review of spectral descriptors found in audio engineering literature, with their corresponding frequency ranges. Descriptors in this list are predominantly associated with equalization effects, and can be characterized by amplification or attenuation applied to a specific band in the magnitude spectrum of a sound.

Similar studies have focused on the elicitation of large datasets of descriptive terms in sound production with respect to audio effect transformations, namely Stables et al. [184], Cartwright et al. [186] and Seetharaman et al. [266]. As the terms in these lists are taken from various effects, we do not necessarily have access to details regarding the corresponding spectral components that are correlated with each descriptor. Instead, we present the distribution of instances of each descriptor across the audio effects that were used to generate them.

Table 9.2 The top 50 Terms from the SAFE [184] Dataset, sorted by number of instances

N	term	Total	Comp	Dist	EQ	Rev
1	warm	582	9	26	542	5
2	bright	531	4	5	521	1
3	punch	34	27	1	6	0
4	room	33	1	0	2	30
5	air	31	0	0	18	13
6	crunch	29	0	27	0	2
7	smooth	22	15	3	2	2
8	vocal	22	16	1	4	1
9	clear	21	3	0	18	0
10	subtle	21	6	4	1	10
11	bass	20	3	4	13	0
12	fuzz	19	1	17	1	0
13	nice	18	12	0	4	2
14	full	16	3	0	9	4
15	boom	15	2	2	9	2
16	crisp	15	1	3	11	0
17	sofa	15	15	0	0	0
18	soft	15	5	1	4	5
19	big	13	1	0	1	11
20	clean	13	1	0	11	1
21	thin	13	1	0	12	0
22	box	12	1	0	8	3
23	deep	12	3	1	6	2
24	tight	12	7	0	4	1
25	drum	11	3	0	2	6
26	gentle	11	6	2	1	2
27	thick	11	2	2	6	1
28	crushed	10	7	2	1	0
29	damp	10	1	1	1	7
30	harsh	10	1	4	5	0
31	low	10	0	0	10	0
32	presence	10	2	0	8	0
33	space	10	0	0	1	9
34	tin	10	0	2	7	1
35	acoustic	9	4	2	3	0
36	comp	9	9	0	0	0
37	dream	9	1	0	0	8
38	flat	9	5	1	3	0
39	hall	9	0	0	0	9
40	kick	9	4	1	4	0
41	loud	9	6	2	1	0
42	present	9	3	0	6	0
43	sharp	9	2	1	4	2
44	small	9	0	0	0	9
45	bite	8	0	0	8	0
46	click	8	1	0	7	0
47	cut	8	2	0	6	0
48	dark	8	0	0	4	4
49	echo	8	0	0	0	8
50	glue	8	8	0	0	0

In the SAFE dataset [81, 184], discussed in detail in Chapter 5, terms are collected via a suite of DAW plugins, comprising a dynamic range compressor, an equalizer, a distortion effect and a reverb effect. This means terms were generated predominantly by expert users, and each term has a corresponding parameter set, an audio feature set and a table of user metadata. The data is sourced from users describing transformations that are made by the DAW plugins when applied to their own audio signals, within their own production workflow. This data has since been used to design Intelligent Music Production tools such as semantically controlled audio effects [374, 443], an audio processing chain generator [128] and a timbral modifier [444].

Similarly, the SocialFX database [185] contains the data from the SocialEQ [186] and Reverbalize [266] projects, along with an additional dynamic range compression study. While the dataset is very large, the entries were crowd-sourced using Amazon's Mechanical Turk platform, suggesting that only a small fraction of the contributors are likely to be expert users. Users contribute terms to the dataset by positioning sounds in a reduced-dimensionality space, which corresponds to audio effect parameters. Like the SAFE dataset, audio effect parameters are available for each instance, along with some metadata extracted from the user. The top 50 terms from the dataset, ranked by number of entries, are presented in Table 9.3.

9.3.2 Term Categorization

As Zacharakis et al. described in [142,379,445,446], descriptive language in music production can be categorized using a number of schemata. This is useful as it allows us to attribute formal meaning to descriptions of sound, and separate potentially context-specific terms such as those associated with an instrument, from terms that represent emotional or musically motivated responses. By differentiating terms in this way, we are able to evaluate the extent to which they are useful under given conditions. For example, it is unlikely that we could use the term *flutey*, a description given to flute samples in a study conducted by Disley et al. [377], to modify the timbre of an electric guitar.

Wake and Asahi [447] proposed a schema to categorize subjective responses to musical stimuli. Three categories were proposed: (1) *onomatopoeia* or *sound itself*, a group used by participants to represent terms that mimic the acoustic sound source, (2) *sound source*, or *sounding situation*, a group representing situational factors of the source (e.g., the instrument or environment) and (3) *adjectives*, or *sound impression*, meaning a subjective, figurative description of the sound.

These groups can be considered sub-categories of a taxonomy proposed by Koelsch [448], in which the author grouped subjective responses to musical sounds using three headings: *extra-musical*, *intra-musical* and *musicogenic* meaning. Extra-musical meaning, also referred to as *designative* meaning, refers to associations between a musical sound source and a non-musical context. Koelsch describes three dimensions of extra-musical meaning, translated from a German study by Karbusicky [449]: (1) *iconic* referring to metaphorical comparisons between the sound and a non-musical quality, (2) *indexical* referring to the expression of an emotional state, and (3) *symbolic* referring to cultural and social references. Most of the semantic terms in music production, including the schema proposed by Wake and Asahi, fall into the *iconic* subcategory. A visual representation of this hierarchical taxonomy is presented in Figure 9.14.

Table 9.3 The top 50 Terms from the SocialFX [185] Dataset, sorted by number of instances

N	term	Total	Comp	EQ	Rev
1	echo	2396	118	0	2278
2	loud	1308	261	21	1026
3	tin	1212	89	28	1095
4	low	1154	92	16	1046
5	war	1137	147	60	930
6	warm	1057	135	59	863
7	church	1033	8	0	1025
8	big	934	55	1	878
9	spacious	855	62	0	793
10	distant	848	29	2	817
11	deep	787	31	6	750
12	muffle	634	85	4	545
13	muffled	623	81	4	538
14	hall	584	7	0	577
15	clear	567	126	8	433
16	ring	537	24	7	506
17	soft	533	102	26	405
18	big-	517	13	0	504
19	bas	506	46	3	457
20	far	473	9	0	464
21	bass	461	43	1	417
22	like	450	9	0	441
23	distort	442	62	0	380
24	nic	432	34	6	392
25	echoing	415	12	0	403
26	the	397	22	1	374
27	large	396	17	2	377
28	-like	377	7	0	370
29	and	361	21	10	330
30	louder	350	156	0	194
31	con	348	13	1	334
32	distorted	343	43	0	300
33	full	337	70	1	266
34	room	332	33	0	299
35	nice	329	30	3	296
36	drum	324	11	1	312
37	hollow	323	14	2	307
38	sad	323	3	21	299
39	high	319	37	4	278
40	strong	316	40	1	275
41	organ	315	0	0	315
42	way	294	8	0	286
43	pleasant	293	32	4	257
44	under	286	7	1	278
45	old	282	18	36	228
46	harp	277	55	8	214
47	smooth	277	13	9	255
48	sound	277	31	1	245
49	metal	270	23	2	245
50	sharp	257	55	7	195

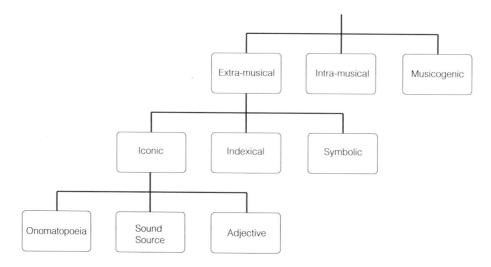

Figure 9.14 The term classification hierarchy, proposed by [447] and [448]

An alternative approach to the categorization of sonic adjectives is proposed by Toulson [450], in which the author proposes that terms fall into the following three categories: (1) *technical*, referring to audio effect parameters of low-level audio features of the signal, (2) *spatial*, referring to the position of a sound in an acoustic environment, or (3) *timbral*, referring to a perceptual dimension of timbre, often described using a metaphor. These groupings are based on a number of studies by Rumsey et al. on the semantics of spatial audio processing [231, 451, 452]. Figure 9.15 illustrates the organization of terms into these groups using the descriptors in [114], some of which are presented in Table 9.1.

9.3.3 Agreement Measures

The extent to which natural language can be used to provide novel methods of modifying sound is determined by the level of exhibited agreement within a term. This concept was investigated by Sarkar et al. [438], in which terms such as *bright, resonant,* and *harsh* all exhibit strong agreement scores, and terms such as *open, hard,* and *heavy* all show low subjective agreement scores. To find this information, participants of a listening test were played randomized samples from the Freesound database[5] and asked to complete a survey attributing descriptive terms to each of the sounds. User agreement towards descriptive terms was found by analyzing the variability in rows of a confusion matrix. This inferred descriptor confidence through labels assigned to samples during repeated trials. In other words, if the same sample is described using the same term by multiple participants, there must be a strong association between the two. When multiple subjects in a study like this use a term for a common purpose, it suggests there is a consensus on its perceptual representation. Terms with a particularly high consensus are useful in Intelligent Music Production as we can utilize them to perform representative modifications of sounds using natural language interfaces.

---

[5] http://freesound.org

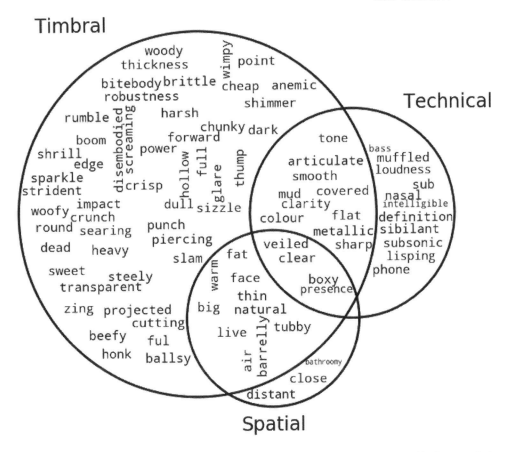

Figure 9.15 A Venn diagram showing the organization of terms into the groups proposed by Rumsey et al.

The notion of inter-subject descriptor agreement was explored further by Cartwright and Pardo [186], in which the agreement for a descriptor d can be considered the overall variance across participants for each of the dimensions in some statistical representation of the term. For example, if we observe all of the audio effect parameters from instances of users trying to make a sound *warmer*, to what extent do each of those parameters vary? When taking the covariance matrix of these statistical parameters, this measurement is equivalent to the trace, as expanded in Equation 9.5

$$trace(\Sigma)_d = \frac{1}{N} \sum_{k=0}^{M} \sum_{n=0}^{N-1} (x_{n,k} - \mu_k)^2 \tag{9.5}$$

In this case, N is the number of instances of the descriptor in the dataset, M is the number of statistical parameters, $x_{n,k}$ is then the kth parameter for instance n and μ_k is the mean of the kth parameter across all of the instances of d. To measure the agreement, the authors also take into consideration the number of entries made into the dataset by dividing the natural log of the number of instances by the trace, as shown in Equation 9.6.

$$A_d = \frac{\ln N}{trace(\Sigma)_d} \qquad (9.6)$$

The parameter variance method has been extended by Stables et al. [81] to cover the distribution of feature values for each term. This allowed to map descriptors from a number of different audio effect transforms and into a common timbre space. The authors used PCA to reduce the dimensionality of a space of consisting of around 100 timbral features (spectral moments, loudness, MFCCs, etc.) per instance, to a 6xN space, from which the trace of the covariance matrix was measured. To evaluate the popularity of each descriptor, this trace was weighted by a coefficient representing the term as a proportion of the dataset, as shown in Equation 9.7.

$$p_d = c_d \times ln\frac{n(d)}{\sum_{d=0}^{D-1} n(d)} \qquad (9.7)$$

where $n(d)$ is the number of entries for descriptor d, and c_d is the output of Equation 9.5, when applied to the reduced dimensionality timbre space.

9.3.4 Similarity Measures

The extent to which descriptions of sound are similar is also of interest to the Intelligent Music Production community, as it allows us to identify synonyms (e.g., "are bright and sharp timbrally equivalent?"), to label parameter scales (e.g., "does something get warmer as it gets less bright?") and to recommend new settings to users based on their current preferences (e.g., "it looks like you're trying to make this sound brighter, try adjusting these parameters for better results"). To measure the similarity of terms, we can use lots of different aspects of the described sounds. Here, we break similarity measures into 2 groups: "timbral similarity", and "contextual similarity".

Timbral Similarity

Timbral similarity measures are based on the use of low-level audio features or audio effect parameters to compute distance or divergence measures between data points or clusters. One of the most commonly used approaches is to apply agglomerative clustering to a dataset of labelled feature sets, and then measure the cophenetic distance (the magnitude of the first common branch) between two data points. This is illustrated in Figure 9.16 using terms taken from the SAFE dataset [184].

Here, the mean of audio feature vectors extracted from each descriptor is computed and PCA is applied. The resulting clusters are intended to retain perceived latent groupings, based on underlying semantic representations. We can see in the EQ data for example (Figure 9.16) that terms generally associated with boosts in high-mid and high frequency bands, such as *tin*, *cut*, *clear*, and *thin* are grouped together, whereas a separate cluster associated with boosts to low and low-mid bands are separated with high cophenetic distance. When we plot these groups, as shown in Figure 9.17, we can see that the spectral profiles of terms within the same cluster are highly correlated. Curves in the first cluster generally exhibit amplification around 500 Hz with a high-frequency roll-off. Similarly, the terms in the second cluster exhibit a boost in high-frequency ($\geq$ 5 kHz) with attenuated low frequencies.

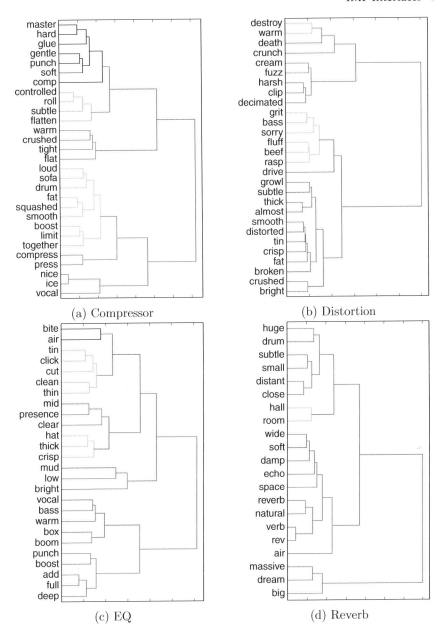

Figure 9.16 Dendrograms showing term clustering based on feature space distances for each transform class (from [81]).

While this form of analysis is predominantly applied to statistical feature-sets, hierarchical clustering is also often applied to data which has been subjectively evaluated by a corpus of expert listeners [453–455]. When this is the case, cophenetic distances represent the similarity of data points, as evaluated by the participants in the study, as opposed to statistical features of the samples.

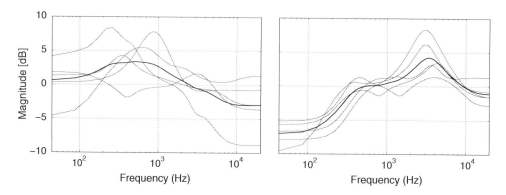

Figure 9.17 Equalization curves for two clusters of terms in the dataset

Contextual Similarity

Contextual similarity measures evaluate the frequency of which a descriptor is entered into a dataset, partitioned by some contextual feature. For example, when terms are derived from musical samples, we can judge the number of entries grouped by attributes such as the genre of the music, the type of instrument being used to perform the sound, or the type of audio effect used to transform the signal. This allows us to evaluate the extent to which terms exhibit similarity without the requirement for feature extraction or storage of low-level parameters.

To identify the way in which terms vary with respect to a given context, we can calculate the *generality* of the term. This can be done by taking the distribution of term-entries across a given context, and then by finding the weighted-mean, as presented in Equation 9.8. This is equivalent to finding the centroid of the density function

$$g = \frac{2}{K-1} \sum_{k=0}^{K-1} k \, \text{sort}(x(d))_k \tag{9.8}$$

where the distribution of the term d is calculated as a proportion of the transform class k to which it belongs.

$$x(d)_k = \frac{n_d(k)}{N(k)} \frac{1}{\sum_{k=0}^{K-1} N(k)} \tag{9.9}$$

Here, $N(k)$ is the total number of entries in class k and $n_d(k)$ is the number of occurrences of descriptor d.

Building on this idea, we can calculate a more comprehensive similarity matrix based on the term's context using techniques taken from natural language processing. In [81], a Vector Space Model (VSM) is used to identify the similarity of each term in the SAFE database, based on the number of entries from each audio effect. First, the term is represented as a 4D vector, where each dimension represents the frequency of entries from a given audio effect, for example $t = [0.0, 0.5, 0.5, 0.0]$ is equally distributed across the compression class and the

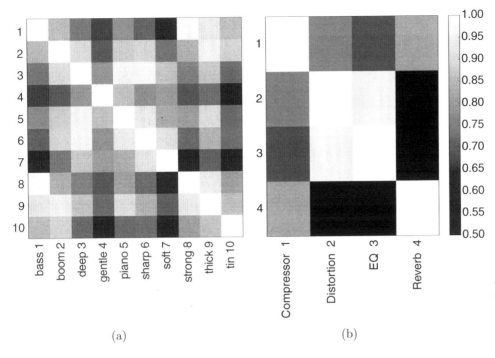

Figure 9.18 Vector-space similarity with regard to (a) high-generality terms and (b) transform-classes

reverb class. Pairwise similarities $(\mathbf{t_1}, \mathbf{t_2})$ are then measured with cosine similarity:

$$\text{simil}(\mathbf{t_1}, \mathbf{t_2}) = \frac{\mathbf{t_1} \cdot \mathbf{t_2}}{\|\mathbf{t_1}\| \|\mathbf{t_2}\|} = \frac{\sum_{i=1}^{N} t_{1,i} t_{2,i}}{\sqrt{\sum_{i=1}^{N} t_{1,i}^2} \sqrt{\sum_{i=1}^{N} t_{2,i}^2}} \qquad (9.10)$$

To better capture true semantic relations of the terms and the transforms they are associated with, latent semantic indexing is applied [456]. This involves reducing the term-transform space from rank four to three by performing a singular value decomposition of the $N_{terms} \times 4$ occurrence matrix $\mathbf{M} = \mathbf{U\Sigma V^*}$, and setting the smallest singular values to zero before reconstructing it using $\mathbf{M'} = \mathbf{U\Sigma'V^*}$. This eliminates noise caused by differences in word usage, for instance due to synonymy and polysemy, whereas the 'latent' semantic relationships between terms and effects are preserved. Figure 9.18 shows the resulting pairwise similarities of the high-generality terms.

Here, the most similar terms are *bass* and *strong*, *deep* and *sharp*, and *boom* and *thick* (all with a cosine similarity of 0.99). Conversely, the similarity of transform types based on their descriptive attributes can be calculated by transposing the occurrence matrix in the VSM. This is illustrated in Figure 9.18, showing terms used to describe equalization transforms are similar to those associated with distortion (0.95), while equalization and compression vocabulary is more disjunct (0.641).

9.3.5 Semantic Audio Processing

Given the potential for the use of natural language as an interface in music production, a number of systems have been developed to map descriptive terminology to processing parameters for both mixing and the application of audio effects. A large number of these models have been designed to allow users to perform equalization using perceptually meaningful terms. An early example of this was the assistive expert system proposed by Reed [9]. The method uses inductive learning, based on nearest neighbor searching of a timbre space. The system allows users to modify the *brightness*, *darkness* and *smoothness* of a sound, by applying a many-to-one mapping to the EQ interface. The author suggests that descriptions are source-dependent, suggesting models for applying semantic processing need to be adaptive to the input signal. This form of semantic equalization was also adopted by the aforementioned SubjEQt [17] and 2DEQ [263] systems. These systems allow users to manipulate features such as *airy*, *nasal* and *muddy* through reduced dimensionality interfaces, constructed with user-data.

A popular technique for the modification of a sound's spectrum using natural language is to identify a number of low-level audio features that are highly correlated with a term or group of terms, and then to develop systems that manipulate the feature directly. In order to modify the perceived brightness of a sound, Brookes and Williams [370] proposed a model for modifying a signal's spectral centroid by applying a spectral tilt filter. This applies gain to each partial as a linear function of its frequency, allowing the spectral centroid to be altered, while preserving the harmonic relationship between spectral components. In similar studies, the authors modified a the relationship between the harmonics of a signal to control perceived warmth [457], and a signal's attack time to control perceived softness [458]. Zacharakis and Reiss [459] also dealt with the manipulation of brightness by altering a signal's spectral centroid. Here, the approach was to split the magnitude spectrum into two subbands, one above and one below the original spectral centroid. The relative amplitudes of these bands can then be altered to manipulate the frequency of the centroid. This preserves the original signal's structure, as no additional spectral components are introduced and the relative levels of components within each band remain the same. Enderby and Stables [444] extended this method using a harmonic distortion algorithm. The spectrum is also split into two bands, one containing low-frequency components, and another containing high frequency harmonics, generated by the nonlinear distortion. A gain parameter from the distortion is then used to control the relative levels, thus manipulating the centroid.

Crowdsourced methods such as the aforementioned SAFE project (proposed by Stables et al. [184]) use templates taken from large datasets of descriptive terms to modify the timbre of a sound using audio effects. The project allows processing to be applied using equalization, distortion, compression and reverb. In each case, terms that have been input by users of the system can be queried, and parameter settings are averaged. These settings are tailored to the user's signal through the use of metadata. The system partitions the data, affording more more specific presets based on external factors such as the sound's genre and instrument. This method has been extended by Stasis et al. [443], by allowing users to navigate specific meanings of each-term (referred to as a sub-representation). Here, clustering is applied a reduced dimensionality representation of the descriptor space, which can then be navigated using an autoencoder.

A similar approach to the manipulation of an audio signal using natural language was followed in the intuitive equalization studies led by Pardo et al. [262, 264, 460]. Algorithms which rapidly learn a desired equalization curve from user data were developed using a number of techniques. Listeners were asked to indicate the extent to which an equalized sound could be described by a perceptual term. After each rating, weights for each frequency band were derived by correlating the gain at each band with listener responses. This provided a mapping from the descriptors to audio processing parameters, leading to the automated construction of an intuitive equalizer UI. This was extended further using active and transfer learning techniques [260]. By exploiting knowledge from prior concepts taught to the system, the performance of the learning algorithm was enhanced.

The SocialEQ [186], Reverbalize [187, 266], and Audealize [461] projects, also by Pardo et al., took a very similar approach to personalization of audio effect parameter spaces. In these cases, users were able to manipulate example sounds with various levels of processing applied, using a low-dimensional interface. This works for either an equalizer or an algorithmic reverb, and allows for visualization of descriptive terms in a 2D space. By defining their own terms, users taught the systems their parameter-space representations through repeat trials. The systems were able to reverse engineer the subject's desired parameter settings using linear regression, leading to an invertible preset. This inversion allows a user to control the amount of processing applied, so they could, for example, control the amount of warmth or brightness of a signal.

10
Future Work

To conclude, this chapter offers a number of key challenges and promising directions in the field of Intelligent Music Production. It is based on the review of the current state of the art, but certainly subjective to a degree. This chapter may therefore inspire a suitable project for students and academics starting to work in this field, or provide a reference for established IMP researchers.

10.1 Conclusions

This book has focused primarily on modeling tasks that a human engineer would have previously been required to carry out. When an engineer makes decisions, they focus not only on the audio content in a single channel, but also on the way that content sounds in the context of a mix. For example, the relationship between sources in a multitrack will usually influence a channel's level, spectrum and dynamics. As will the genre of the music, or the year in which is it produced. This means that even for single track, monaural production tasks, the challenge of automating processing is not necessarily tractable. Further to this, the large number of tasks that are embedded in the recording, mixing and mastering processes mean there is a requirement for a range of diverse algorithmic tools, that can often conflict with each other, and are dependent on the success of the modules that are executed before them in the processing chain.

This complexity often leads to the requirement for human input to intelligent functions, and means that a large section of IMP systems aim to tackle smaller, more fundamental tasks in audio engineering. Automatic mixing for example tends to focus on optimizing intelligibility across sound sources, as opposed to producing stylized release-ready outputs. Similarly, cross-adaptive processing techniques such as multitrack equalization focus on reducing statistically quantifiable features such as auditory masking, as opposed to emulating the way in which a mixing engineer would manipulate timbre. This highlights one of the key shortcomings of today's IMP systems, and will likely form the basis of future iterations of intelligent systems in audio engineering, which is that there is still a vast disparity between computational models of audio engineers, and real, human audio engineers.

10.2 What We Have Covered

The book has been presented in three sections. In the first section, we started by discussing the things we, as audio engineers, already know about music production. We first summarized our understanding of audio effects, based on the ways in which they are classified and applied to musical signals. The fact that audio effects can be categorized based on a number of their attributes demonstrates the complexity of selection by a human engineer. We introduced a number of control mechanisms, and showed that adaptivity can manipulate parameters using not only the source being processed, but also other sources in a mix. In Chapter 3, we then focused on the skills an engineer has, and some of the assumptions that are made by humans when mixing. We summarized the challenges of mixing, and discuss key variables such as the acoustic environment, the style of music, and the subjective input of the engineer. We showed through extensive research that we are able to find some generalized rules, though they are often conflicting and context-dependent. We presented some of the key challenges in emulating human engineers, including the role of external factors such as the location and experience level of the engineer, the style of the music, and the approach a given studio has to mixing.

The second section of the book focused on the development of IMP building blocks. To build fully functional models, we addressed the requirements of a range of intelligent systems and their components. Chapter 4 specifically addressed the construction of systems for automatic mixing and explored some of the more common parameter configurations in IMP system design. We showed the importance of feature extraction and side-chain processing, and reviewed the state of the art in automatic mixing systems, which can be categorized by their underlying mechanisms. For example, some use heuristic rules and knowledge engineering, whereas some use machine learning to complete the same tasks. In Chapter 5, we discussed the availability and importance of high-quality datasets, and provided a range of examples. We mainly focused on multitrack audio content, with varying levels of annotation and metadata, as this is the key driver behind multichannel processing tools like cross-adaptive effects and automatic mixing algorithms. Chapter 6 summarized methods for subjective evaluation of IMP systems. This is particularly important given the fact most of the mixing process is undertaken explicitly to make subjectively aesthetic improvements to the audio. Here, we present testing methodologies such as MUSHRA, and discuss a number of best practices.

The final section of the book put the pieces together in order to present fully formed IMP tools. Chapter 7 addressed IMP systems, in which the focus is on adding intelligence to a range of effects such as panning, equalization, and dynamics processing. For each effect we presented a number of assumptions relating to the ways in which it should be used, followed by methods for automating its parameters. Chapter 8 addressed intelligent processes, i.e., the automation of tasks which are often performed by an engineer such as track editing or channel routing. We showed that a large number of tasks can be made more intuitive, or can be taken care of with very little input from an engineer. Finally, in Chapter 9 we presented the user interfaces designed through abstract representation, or data-driven analysis. We provided a number of alternatives to the channel-strip interface, which has been constant since the early days of audio engineering. We finished by exploring other mechanisms for controlling audio processors, such as using natural language to describe parameter settings.

10.3 Future Directions

Intelligent Music Production is still a relatively new field, so there are a number of exciting research directions. Many of the systems presented throughout this book are still in their infancy due to a number of factors, be it processing power, the availability of data, or the current state of the art. This leaves ample room for improvement on existing models. In the past few years alone, researchers from different disciplines have targeted all aspects of the mixing process, using a plethora of techniques from statistical and analytical data-driven processing, to heuristic rule-based processing. Here, we share a small number of the directions we feel are going to be prominent in the next few years, and will shape the future of IMP.

10.3.1 Data Availability

A significant obstacle to the development of high-quality systems, especially those based on machine learning, is the relative shortage of reliable data to inform or test assumptions about mix practices [3]. Recent efforts towards sharing datasets (see Chapter 5) and accommodating efficient capture of mix actions [120] may help produce the critical mass of data needed for truly intelligent mixing systems. With analysis of high volumes of data it may also become possible to uncover the rules that govern not just mix engineering in general, but particular mixing styles [114]. From an application point of view, a target profile can thus be applied to source content to mimic the approach of a certain engineer [29], to fit a specific musical genre, or to achieve the most suitable properties for a given medium. Taking the target approach of specific effects described in Chapter 7 to the overall mix level, one can take two completely different mixes and try to impose the sound characteristics of one onto the other. This concept of style transfer is being explored in depth in other areas of the arts including image processing [462] and musical performance [463] and, with the advent of new deep learning systems, has seen a great deal of success.

Different genres seem to require different treatments too, such as a higher drum loudness in rock music compared to jazz [139]. This distinction presents an obstacle for one-size-fits-all music production systems, but offers opportunities for high-level user interaction, such as the idea of an 'InstaMix' filter (see Section 1.3.1). To accommodate this variety, research in this area needs to employ datasets of sufficient size and diversity to determine the impact of genre. The established mixing rules also need to incorporate this information. As data mining and machine learning continue to permeate all aspects of technology, including music production, this may soon become a reality. The music industry has the potential to collect huge amounts of user generated data, and to gather music consumption information such as skip rate, number of plays, adds to playlists, etc. Some of this information is already being collected on a large scale by music streaming platforms. This type of data could be used to track automatically produced content, and then adapt the processing to optimize such values. Furthermore, artificial intelligence systems are beginning to understand human subjectivity, which will lead to their ability to address the aesthetic aspects of a mix.

10.3.2 Multitrack Audio Effects

Multitrack audio effects are relatively rare in mainstream music production due to a number of limitations imposed by modern workstations and plugin architectures. This means effects are essentially single track, with single input and output sources. Although frameworks such as VST3 and JSAP [201] support cross-adaptive digital audio effects, very few software-based mixers allow multitrack plugins. Furthermore, inserts are pre- or post-fader, thus preventing effective use of intelligent, automatic faders and equalizers. This limits a plugin's access points to other channel-data in the mix.

Due to the heavy reliance on side-chain processing, we currently make use of input audio and feature data from other channels being processed in parallel at the same point in the timeline. As frameworks become more open, these systems may perform more accurately if they have access to features extracted from channel outputs as well as from the processed inputs. Similarly, if functions have access to audio at various points along the timeline, we may be able to make more comprehensive predictive models that take the structure of the composition into account, and vary production parameters accordingly.

In order to address this lack of multitrack processing, audio workstations need to adapt to support cross-adaptive effects. An initial eight channel software prototype of such a system was built for the evaluation of intelligent multitrack signal processing tools described in [151], and integrated with a Mackie hardware control surface for demonstration. In [46, 49], cross-adaptive VST plugins with support for up to 64 tracks were custom-built, and used in Reaper. However, none of these demonstrators were intended to support additional cross-adaptive effects beyond those that were being developed. Any input or output stage in the automatic mixing host should have the capability to be analyzed, processed, stored and recalled. This would allow for portability and consistency. Hardware implementations would permit a host system to be tested in live situations. And hardware could be evolved from an automated mixer to an intelligent, automatic mixer.

10.3.3 Mixing for Hearing Loss

Almost all discussion herein has assumed that all listeners have the same hearing abilities. But a very large portion of the population have some form of hearing loss. It may be possible to adapt a mix so that it might be perceived the same by someone with different hearing capabilities [464], or mix specifically for cochlear implants [465, 466]. This would increase accessibility in music production, and facilitate some of the haptic and visual tools presented in Chapter 9, opening up production to a wider range of users.

10.3.4 Interactive Audio

Auto-spatialization

Few best practices are known about spatial positioning for object-based audio, or any formats beyond stereo, if they exist at all. Given the increasing popularity of virtual reality applications and other immersive technologies, guidelines for intelligent music production in this area will become crucial in the coming years. Partly due to this lack of knowledge, autonomous spatial positioning of sources has been limited to stereo panning. Object tracking improvement would permit automatic spatialization beyond stereo sources into other spatial audio formats like ambisonics, 5.1 surround or vector based amplitude panning (VBAP) [467]. This would allow Automatic Mixing applications in the increasingly important domain of VR systems, as well as game and film audio. The use of tracking devices for determining the musician's position could also be used for performing rescaling and thus as a form of maintaining system stability and avoiding unwanted acoustic feedback artifacts. Tracking could be performed using bluetooth or wireless microphones. As both devices use radio frequency transmission, theoretically the location can be estimated using power triangulation.

Remixing Formats

Fueled by increasing popularity of object-based audio, sound can be mixed based on metadata, environmental parameters and user preferences. In such scenarios, the medium would carry stems rather than a full mix, or source separation would be performed at the consumer side. Possible applications include adjusting parameters depending on location or activity, adapting reverberation to the listening environment or full 're-mixing' functionality for the end-user. This type of processing is supported by formats discussed in Chapter 5 such as Native Instruments' Stems, in which multiple streams can be encapsulated in an MP4 file.

Game Audio

Game audio production presents many challenges [468]. In traditional linear media, the sound engineer knows in advance the sequence and combination of audio events. But for interactive media, events occur in response to actions by the user. The audio is often preprocessed for many game scenarios, and submixes are created manually in order to ensure that in-game mixing is kept simple. However, this preprocessing results in constraints on the ability to mix in the game. Furthermore, excessive use of preprocessed audio results in sound repetition, leading to listener fatigue.

Alternatively, some sound processing is managed by an in-game audio engine. But without the ability to perform signal analysis, and only limited knowledge of the game scenario, in-game mixing is limited. For individual audio tracks, the sound designer may provide parameters which are used by the audio engine to guide the manipulation of content. For instance, the sound designer may set up a volume-vs.-distance curve, specifying how the volume of the dialog track should be attenuated based on the virtual distance from speaker to player. This information is then used by the in-game audio mixing engine to guide the processing of audio content upon use in the game. However, this metadata is specific to each audio track, independent of the others, and hence does not exploit the relationships between tracks. In-game, multitrack mixing is usually limited to adjusting the gains of concurrent tracks based on their priority [469], and no attempt has been made to use equalization, repositioning or delaying of sources to ensure intelligibility and audibility while emphasizing the most relevant content.

Dynamic game mixing obeys the same rules as automatic mixing for live sound, but it is driven by game scenarios. As with live sound, game audio should be edited in real time and the sound designer or engineer does not have a priori knowledge of the signal content. In live sound, room acoustics dominate. But in game sound, the physics of the scene is only a starting point. The sound designer dictates relevance and impact, and the player acts as a moving directional microphone with flexible point of view. Hence, priorities are assigned to sound spots or cues, and automatic mixing is performed with weighted constraints and overrides. So our goal is to have three tiers of automatic mixing: game-relevant stem production, data driven flexible mixing and situation-driven in-game mixing. To achieve this we have layers of fuzzy constraints where the game state dictates limits on the panning, delays, equalization and gain that may be applied in order to minimize masking while bringing selected sounds to the foreground.

The goal would be to develop intelligent tools for interactive audio, where listener actions and context determine how the audio should be mixed. A significant challenge would be to adapt our Intelligent Systems approach to variable, game-specific constraints, i.e., the best sound rendering need not be realistic and may change depending on the game situation.

10.3.5 Turing Test

Ideally, intelligent systems for mixing multitrack audio should be able to pass a Turing test [80]. That is, they should be able to produce music indistinguishable from that which could be handcrafted by a professional human engineer. Nearly ten years ago, an automatically generated mix was submitted to an Audio Engineering Society recording competition, disguised as a human-made entry [151]. The judges commented it was far from what was expected from a professional engineer, but were surprised when told it was machine-generated. There is no doubt that the current state of the art could produce significantly better results on a wider variety of content. Still, consistently performing on par with professional engineers would require the systems to be able to make excellent artistic as well as technical decisions, and be ready to expect almost arbitrary audio content. Considerable progress is also needed in order for systems to be able to 'understand' the musician's intent.

10.4 Final Thoughts

In this book, we described how music production could be made simpler and more efficient through the use of intelligent software tools. In the not-so-distant future, this may result in several classes of systems, with varying degrees of user control. The first would be a set of tools for the sound engineer which automate repetitive tasks. This would free up professional audio engineers to focus on the creative aspects of their craft, and help inexperienced users create high-quality mixes. Others would take the form of a 'black box' for musicians which allows decent live sound without an engineer. This would be most beneficial for the small band or small venue that don't have or can't afford a sound engineer, or for recording practice sessions where a sound engineer is not typically available. Still other systems would not so much take control, but fundamentally change the way of control on an interface or workflow level, making abstraction from the underlying implementation and fully leveraging today's technologies and computing power.

Since this field emerged, there have been concerns with its premise. If full automation is successful, then machines may replace sound engineers. But much of what a sound engineer does is creative, and based on artistic decisions. It is doubtful that such decisions would be definitively in the hands of a machine, much like other arts and trades have been transformed through technology, but not disappeared. These classes of tools are certainly not intended to remove the creativity from audio production. Rather, the tools rely on the fact that many of the challenges are technical engineering tasks. Some of these are perceived as creative decisions because there are a wide range of approaches without a clear understanding of listener preferences. Automating those engineering aspects of record production and sound reinforcement will allow the musicians to concentrate on the music and the audio engineers to concentrate on the more interesting, creative challenges.

Appendix A
Additional Resources

Intelligent Sound Engineering Blog

Musings about music production, psychoacoustics, sound synthesis and related topics, from the Centre for Digital Music's Audio Engineering research team. https://intelligentsoundengineering.wordpress.com/

Intelligent Sound Engineering YouTube Channel

The YouTube channel of the Audio Engineering team at Queen Mary University of London https://youtu.be/user/IntelligentSoundEng. It contains videos of many automatic mixing systems based on multitrack signal processing, including;

- Combined automatic mixer https://youtu.be/ZKrCNQC7HW4
- Spectral enhancer https://youtu.be/gES8cnAfHHg
- Automatic gain controls https://youtu.be/mUwt0Kf5nv0
- Gain normalization https://youtu.be/oJLD-lXQtNg https://youtu.be/NIOVO9Ho_LI
- Bleed and interference removal https://youtu.be/4M536OsoK6U https://youtu.be/uC2YU_Stk0Q
- Comb filter prevention https://youtu.be/ieOwP-vN5Uk https://youtu.be/uFEvyCqSEsM https://youtu.be/FPq63RqJZZY
- Reverse engineering the mix https://youtu.be/watch?v=mC95mSEXwRA
- Time offset and polarity correction https://youtu.be/watch?v=4OxkHGaNerQ
- Interactive mixing https://youtu.be/Dp0N0HS3L2s
- Autofaders https://youtu.be/b-02Ts7RO9M
- Autopanning https://youtu.be/AZ753ni4fjw https://youtu.be/OzcQ5Bajx80 https://youtu.be/1Y9UDAAuZ3M
- AutoEQ https://youtu.be/LYLZYBVLOs8 https://youtu.be/I-u-xIFZd40 https://youtu.be/kdo5vOySG1E
- Adaptive distortion https://youtu.be/gTYYbEqFu_I
- Intelligent dynamic range compression https://youtu.be/8dgguI4A5gA

Interactive Audio Lab

http://music.cs.northwestern.edu The YouTube channel of Bryan Pardo's research team at Northwestern University has created excellent new music production interfaces, as well as crowdsourcing vast amounts of data on how people describe audio effects and attributes. Their YouTube channel https://youtu.be/channel/UCY-jggB_-R3rYaTsWN2_Ssw contains some excellent demonstrations of their work;

- SocialEQ https://youtu.be/Oz3b-IC56F4 with the online demonstration at http://socialeq.org/socialeq/SocialEQ.html
- Mixploration https://youtu.be/qix2nOQ3z5A
- Reverbalize https://youtu.be/SprtYHuX_4g with the online demonstration at https://reverbalize.appspot.com

Web Audio Evaluation Tool

A tool based on the Web Audio API to perform perceptual audio evaluation tests in the web browser, locally or on remote machines over the web https://github.com/BrechtDeMan/WebAudioEvaluationTool, and a video explanation at https://youtu.be/aHmgSSVaPRY

The SAFE Dataset

Discussed in Section 5.2.1, this consists of audio engineering parameter settings, with matching descriptions and audio features [184]. The dataset was collected in order to specifically understand the relationship between language and audio effects www.semanticaudio.co.uk

Stems

This is an open multichannel format, which encodes four separate streams containing different musical elements in a modified MP4 file format www.native-instruments.com/en/specials/stems

HDF5

A framework for storing audio metadata, this allows cross-platform extraction, with fast data I/O and heterogeneous data storage. www.hdfgroup.org/solutions/hdf5

Table A.1 Spectral descriptors in practical sound engineering literature

Term		Range	Reference
air[1]		5–8 kHz	[127, p. 119]
		10–20 kHz	[439, p. 99]
		10–20 kHz	[108, p. 211]
		11–22.5 kHz	[127, p. 26]
		12–15 kHz	[440, p. 103]
		12–16 kHz	[115, p. 43]
		12–20 kHz	[439, p. 25]
		12–20 kHz	[441, p. 108]
		12–20 kHz	[442, p. 86]
anemic	lack of	20–110 Hz	[108, p. 211]
	lack of	40–200 Hz	[127, p. 119]
articulate		800–5000 Hz	[127, p. 119]
ballsy		40–200 Hz	[127, p. 119]
barrelly		200–800 Hz	[127, p. 119]
bathroomy		800–5000 Hz	[127, p. 119]
beefy		40–200 Hz	[127, p. 119]
big		40–250 Hz	[127, p. 25]
bite		2–6 kHz	[441, p. 106]
		2.5 kHz	[285, p. 484]
body		100–500 Hz	[439, p. 99]
		100–500 Hz	[108, p. 211]
		150–600 Hz	[439, p. 24]
		200–800 Hz	[127, p. 119]
		240 Hz	[285, p. 484]
boom(y)		20–100 Hz	[108, p. 211]
		40–200 Hz	[127, p. 119]
		60–250 Hz	[127, p. 25]
		62–125 Hz	[115, p. 43]
		90–175 Hz	[127, p. 26]
		200–240 kHz	[285, p. 484]
bottom		40–100 Hz	[127, p. 119]
		45–90 Hz	[127, p. 26]
		60–120 Hz	[285, p. 484]
		62–300 Hz	[115, p. 43][2]
boxy, boxiness		250–800 Hz	[108, p. 211]
		300–600 Hz	[127, p. 31]
		300–900 Hz	[115, p. 43]
		800–5000 Hz	[127, p. 119]
bright		2–12 kHz	[115, p. 43]
		2–20 kHz	[108, p. 211]
		5–8 kHz	[127, p. 119]
brilliant, brilliance		5–8 kHz	[127, p. 119]
		5–11 kHz	[108, p. 211]
		5–20 kHz	[285, p. 484]
		6–16 kHz	[127, p. 25]
brittle		5–20 kHz	[285, p. 484]
		6–20 kHz	[439, p. 25]

[1] In some books, 'air' is also used to denote a part of the audible frequency range, exceeding 'highs' [442, p. 86], [441, p. 108].
[2] More specifically, [115] calls this 'extended bottom.'

Table A.1 (Cont.)

Term		Range	Reference
cheap	lack of	8–12 kHz	[127, p. 119]
chunky		800–5000 Hz	[127, p. 119]
clarity		2.5–4 kHz	[442, p. 86]
		2.5–5 kHz	[285, p. 484]
		3–12 kHz	[108, p. 211]
		4–16 kHz	[127, p. 26]
clear		5–8 kHz	[127, p. 119]
close		2–4 kHz	[285, p. 484]
		4–6 kHz	[127, p. 25]
colour		80–1000 Hz	[108, p. 211]
covered	lack of	800–5000 Hz	[127, p. 119]
crisp, crispness		3–12 kHz	[108, p. 211]
		5–10 kHz	[285, p. 484]
		5–12 kHz	[127, p. 119]
crunch		200–800 Hz	[127, p. 119]
		1400–2800 Hz	[127, p. 26]
cutting		5–8 kHz	[127, p. 119]
dark	lack of	5–8 kHz	[127, p. 119]
dead	lack of	5–8 kHz	[127, p. 119]
definition		2–6 kHz	[441, p. 106]
		2–7 kHz	[108, p. 211]
		6–12 kHz	[127, p. 26]
disembodied		200–800 Hz	[127, p. 119]
distant	lack of	200–800 Hz	[127, p. 119]
	lack of	700–20 000 Hz	[108, p. 211]
	lack of	4–6 kHz	[127, p. 25]
	lack of	5 kHz	[285, p. 484]
dull	lack of	4–20 kHz	[108, p. 211]
	lack of	5–8 kHz	[127, p. 119]
	lack of	6–16 kHz	[115, p. 43]
edge, edgy		800–5000 Hz	[127, p. 119]
		1–8 kHz	[108, p. 211]
		3–6 kHz	[127, p. 26]
		4–8 kHz	[115, p. 43]
fat		50–250 Hz	[108, p. 211]
		60–250 Hz	[127, p. 25]
		62–125 Hz	[115, p. 43]
		200–800 Hz	[127, p. 119]
		240 Hz	[285, p. 484]
flat	lack of	8–12 kHz	[127, p. 119]
forward		800–5000 Hz	[127, p. 119]
full(ness)		40–200 Hz	[127, p. 119]
		80–240 Hz	[285, p. 484]
		100–500 Hz	[439, p. 99]
		175–350 Hz	[127, p. 26]
		250–350 Hz	[115, p. 43]

Table A.1 (Cont.)

Term		Range	Reference
glare		8–12 kHz	[127, p. 119]
glassy		8–12 kHz	[127, p. 119]
harsh		2–10 kHz	[108, p. 211]
		2–12 kHz	[115, p. 43]
		5–20 kHz	[285, p. 484]
heavy		40–200 Hz	[127, p. 119]
hollow	lack of	200–800 Hz	[127, p. 119]
honk(y)		350–700 Hz	[127, p. 26]
		400–3000 Hz	[108, p. 211]
		600–1500 Hz	[439, p. 24]
		800–5000 Hz	[127, p. 119]
horn-like		500–1000 Hz	[285, p. 484]
		500–1000 Hz	[127, p. 25]
		800–5000 Hz	[127, p. 119]
impact		62–400 Hz	[115, p. 43]
intelligible		800–5000 Hz	[127, p. 119]
		2–4 kHz	[285, p. 484]
in-your-face		1.5–6 kHz	[439, p. 24]
lisping		2–4 kHz	[127, p. 25]
live		5–8 kHz	[127, p. 119]
loudness		2.5–6 kHz	[108, p. 211]
		5 kHz	[285, p. 484]
metallic		5–8 kHz	[127, p. 119]
mud(dy)		16–60 Hz	[127, p. 26]
		20–400 Hz	[108, p. 211]
		60–500 Hz	[441, p. 104]
		150–600 Hz	[439, p. 24]
		175–350 Hz	[127, p. 26]
		200–400 Hz	[115, p. 43]
		200–800 Hz	[127, p. 119]
muffled	lack of	800–5000 Hz	[127, p. 119]
nasal		400–2500 Hz	[108, p. 211]
		500–1000 Hz	[441, p. 105]
		700–1200 Hz	[115, p. 43]
		800–5000 Hz	[127, p. 119]
natural tone		80–400 Hz	[108, p. 211]
oomph		150–600 Hz	[439, p. 24]
phonelike		800–5000 Hz	[127, p. 119]
piercing		5–8 kHz	[127, p. 119]
point		1–4 kHz	[127, p. 27]
power(ful)		16–60 Hz	[127, p. 26]
		40–200 Hz	[127, p. 119]
		40–100 Hz	[108, p. 211]
presence		800–12k Hz	[127, p. 119]
		1.5–6 kHz	[439, p. 24]
		2–8 kHz	[115, p. 43]

Table A.1 (Cont.)

Term	Range	Reference
	2–11 kHz	[108, p. 211]
	2.5–5 kHz	[285, p. 484]
	4–6 kHz	[127, p. 25]
projected	800–5000 Hz	[127, p. 119]
punch	40–200 Hz	[127, p. 119]
	62–250 Hz	[115, p. 43][3]
robustness	200–800 Hz	[127, p. 119]
round	40–200 Hz	[127, p. 119]
rumble	20–100 Hz	[108, p. 211]
	40–200 Hz	[127, p. 119]
screamin'	5–12 kHz	[127, p. 119]
searing	8–12 kHz	[127, p. 119]
sharp	8–12 kHz	[127, p. 119]
shimmer	7.5–12 kHz	[285, p. 484]
shrill	5–7.5 kHz	[285, p. 484]
	5–8 kHz	[127, p. 119]
sibilant, sibilance	2–8 kHz	[108, p. 211]
	2–10 kHz	[115, p. 43]
	4 kHz	[441, p. 120]
	5–20 kHz	[285, p. 484]
	6–12 kHz	[127, p. 26]
	6–16 kHz	[127, p. 25]
sizzle, sizzly	6–20 kHz	[108, p. 211]
	7–12 kHz	[441, p. 107]
	8–12 kHz	[127, p. 119]
slam	62–200 Hz	[115, p. 43]
smooth	5–8 kHz	[127, p. 119]
solid(ity)	35–200 Hz	[108, p. 211]
	40–200 Hz	[127, p. 119]
	62–250 Hz	[115, p. 43]
sparkle, sparkling	5–10 kHz	[127, p. 27]
	5–15 kHz	[108, p. 211]
	5–20 kHz	[285, p. 484]
	8–12 kHz	[127, p. 119]
steely	5–8 kHz	[127, p. 119]
strident	5–8 kHz	[127, p. 119]
sub-bass	16–60 Hz	[127, p. 25]
subsonic	0–20 Hz	[108, p. 209]
	0–25 Hz	[442, p. 84]
	10–60 Hz	[441, p. 102]
sweet	250–400 Hz	[115, p. 43]
	250–2000 Hz	[127, p. 25]
thickness	20–500 Hz	[108, p. 211]
	40–200 Hz	[127, p. 119]
	200–750 Hz	[115, p. 43]

[3] More specifically, [115] calls this 'punchy bass.'

Table A.1 (Cont.)

Term		Range	Reference
thin	lack of	20–200 Hz	[108, p. 211]
	lack of	40–200 Hz	[127, p. 119]
	lack of	60–250 Hz	[127, p. 25]
	lack of	62–600 Hz	[115, p. 43]
thump		40–200 Hz	[127, p. 119]
		90–175 Hz	[127, p. 26]
tinny		1–2 kHz	[285, p. 484]
		1–2 kHz	[127, p. 25]
		5–8 kHz	[127, p. 119]
tone		500–1000 Hz	[441, p. 105]
transparent	lack of	4–6 kHz	[127, p. 25]
tubby		200–800 Hz	[127, p. 119]
veiled	lack of	800–5000 Hz	[127, p. 119]
warm, warmth		90–175 Hz	[127, p. 26]
		100–600 Hz	[108, p. 211]
		200 Hz	[285, p. 484]
		200–800 Hz	[127, p. 119]
		200–500 Hz	[441, p. 105]
		250–600 Hz	[115, p. 43]
whack		700–1400 Hz	[127, p. 26]
wimpy	lack of	40–200 Hz	[127, p. 119]
	lack of	55–500 Hz	[108, p. 211]
woody		800–5000 Hz	[127, p. 119]
woofy		800–5000 Hz	[127, p. 119]
zing		4–10 kHz	[439, p. 99]
		10–12 kHz	[439, p. 24]
bass/low end/		25–120 Hz	[442, p. 84]
lows		20–150 Hz	[439, p. 23]
		20–250 Hz	[108, p. 209][4]
		20–250 Hz	[115, p. 43]
		20–250 Hz	[470, p. 72]
		40–200 Hz	[127, p. 119]
		60–150 Hz	[441, p. 103]
		60–250 Hz	[127, p. 25]
low mids/		120–400 Hz	[442, p. 85]
lower midrange		150–600 Hz	[439, p. 24]
		200–500 Hz	[441, p. 104]
		250–500 Hz	[115, p. 43]
		200–800 Hz	[127, p. 119]
		250–1000 Hz	[470, p. 73]
		250–2000 Hz	[127, p. 25]
		250–2000 Hz	[108, p. 209]
(high) mids/		250–6000 Hz	[115, p. 43][5]

[4] [108] distinguishes between low bass (20–60 Hz), mid bass (60–120 Hz) and upper bass (120–250 Hz)
[5] [115] distinguishes between lower midrange (250–500 Hz), midrange (250–2000 Hz) and upper midrange (2–6 kHz).

Table A.1 (Cont.)

Term	Range	Reference
upper midrange	350–8000 Hz	[442, p. 85][6]
	600–1500 Hz	[439, p. 24]
	800–5000 Hz	[127, p. 119]
	1–10 kHz	[470, p. 73]
	1.5–6 kHz	[439, p. 24]
	2–4 kHz	[127, p. 25]
	2–6 kHz	[108, p. 209]
	2–6 kHz	[441, p. 106]
highs/high end/ treble	5–12 kHz	[127, p. 119][7]
	6–20 kHz	[108, p. 209]
	6–20 kHz	[439, p. 24]
	6–20 kHz	[115, p. 43][8]
	7–12 kHz	[441, p. 107]
	8–12 kHz	[442, p. 86]
	10–20 kHz	[470, p. 74]

[6] [442] distinguishes between midrange (350–2000 Hz) and upper midrange (2–8 kHz).
[7] [127] distinguishes between highs (5–8 kHz) and super highs (8–12 kHz)
[8] [115] distinguishes between lower treble or highs (6–12 kHz) and extreme treble (12–20 kHz).

Bibliography

[1] G. Papadopoulos and G. Wiggins, "AI methods for algorithmic composition: A survey, a critical view and future prospects," in *AISB Symposium on Musical Creativity*, 1999.

[2] V. Verfaille, M. M. Wanderley, and P. Depalle, "Mapping strategies for gestural and adaptive control of digital audio effects," *Journal of New Music Research*, vol. 35, no. 1, pp. 71–93, 2006.

[3] B. Kolasinski, "A framework for automatic mixing using timbral similarity measures and genetic optimization," in *Audio Engineering Society Convention 124*, 2008.

[4] G. Bocko, M. Bocko, D. Headlam, J. Lundberg, and G. Ren, "Automatic music production system employing probabilistic expert systems," in *Audio Engineering Society Convention 129*, 2010.

[5] A. T. Sabin and B. Pardo, "2DEQ: an intuitive audio equalizer," in *7th ACM Conference on Creativity and Cognition*, 2009.

[6] G. Bromham, "How can academic practice inform mix-craft?," in *Mixing Music*, pp. 245–246, Routledge, 2017.

[7] R. Toulson, "Can we fix it? – The consequences of 'fixing it in the mix' with common equalisation techniques are scientifically evaluated," *Journal of the Art of Record Production*, vol. 3, 2008.

[8] A. Pras, C. Guastavino, and M. Lavoie, "The impact of technological advances on recording studio practices," *Journal of the American Society for Information Science and Technology*, vol. 64, no. 3, pp. 612–626, 2013.

[9] D. Reed, "A perceptual assistant to do sound equalization," in *5th International Conference on Intelligent User Interfaces*, 2000.

[10] A. Bell, E. Hein, and J. Ratcliffe, "Beyond skeuomorphism: The evolution of music production software user interface metaphors," *Journal on the Art of Record Production*, vol. 9, 2015.

[11] C. Liem, A. Rauber, T. Lidy, R. Lewis, C. Raphael, J. D. Reiss, T. Crawford, and A. Hanjalic, "Music information technology and professional stakeholder audiences: Mind the adoption gap," in *Dagstuhl Follow-Ups*, vol. 3, Schloss Dagstuhl-Leibniz-Zentrum fuer Informatik, 2012.

[12] P. D. Pestana and J. D. Reiss, "Intelligent audio production strategies informed by best practices," in *Audio Engineering Society Conference: 53rd International Conference: Semantic Audio*, 2014.

[13] P. White, "Automation for the people," *Sound on Sound*, vol. 23, no. 12, 2008.

[14] D. T. N. Williamson, "Design of tone controls and auxiliary gramophone circuits," *Wireless World*, vol. 55, pp. 20–29, 1949.

[15] P. J. Baxandall, "Negative-feedback tone control," *Wireless World*, vol. 58, pp. 402–405, 1952.

[16] G. Massenburg, "Parametric equalization," in *Audio Engineering Society Convention 42*, 1972.

[17] S. Mecklenburg and J. Loviscach, "subjEQt: Controlling an equalizer through subjective terms," in *Human Factors in Computing Systems (CHI-06)*, 2006.

[18] V. Valimaki and J. D. Reiss, "All about equalization: Solutions and frontiers," *Applied Sciences*, vol. 6, no. 5, 2016.

[19] L. B. Tyler, "An above threshold compressor with one control," in *Audio Engineering Society Convention 63*, 1979.

[20] G. W. McNally, "Dynamic range control of digital audio signals," *Journal of the Audio Engineering Society*, vol. 32, no. 5, pp. 316–327, 1984.

[21] A. T. Schneider and J. V. Hanson, "An adaptive dynamic range controller for digital audio," in *IEEE Pacific Rim Conference on Communications, Computers and Signal Processing*, 1991.

[22] E. Vickers, "Automatic long-term loudness and dynamics matching," in *Audio Engineering Society Convention 111*, 2001.

[23] R. J. Cassidy, "Level detection tunings and techniques for the dynamic range compression of audio signals," in *Audio Engineering Society Convention 117*, 2004.

[24] D. Dugan, "Automatic microphone mixing," *Journal of the Audio Engineering Society*, vol. 23, no. 6, pp. 442–449, 1975.

[25] G. Ballou, *Handbook for Sound Engineers*. Taylor & Francis, 2013.

[26] D. Dugan, "Tutorial: Application of automatic mixing techniques to audio consoles," *SMPTE Journal*, vol. 101, no. 1, pp. 19–27, 1992.

[27] S. Julstrom and T. Tichy, "Direction-sensitive gating: a new approach to automatic mixing," *Journal of the Audio Engineering Society*, vol. 32, no. 7/8, pp. 490–506, 1984.

[28] F. Pachet and O. Delerue, "On-the-fly multi-track mixing," in *Audio Engineering Society Convention 109*, 2000.

[29] H. Katayose, A. Yatsui, and M. Goto, "A mix-down assistant interface with reuse of examples," in *International Conference on Automated Production of Cross Media Content for Multi-Channel Distribution*, 2005.

[30] R. B. Dannenberg, "An intelligent multi-track audio editor," in *International Computer Music Conference*, 2007.

[31] Y. Liu, R. B. Dannenberg, and L. Cai, "The Intelligent Music Editor: Towards an Automated Platform for Music Analysis and Editing," in *Advanced Intelligent Computing Theories and Applications Conference*, 2010.

[32] N. Tsingos, E. Gallo, and G. Drettakis, "Perceptual audio rendering of complex virtual environments," *ACM Transactions on Graphics (Proceedings of SIGGRAPH-04)*, vol. 23, no. 3, 2004.

[33] N. Tsingos, "Scalable perceptual mixing and filtering of audio signals using an augmented spectral representation," in *8th International Conference on Digital Audio Effects (DAFx-05)*, 2005.

[34] P. Kleczkowski, "Selective mixing of sounds," in *Audio Engineering Society Convention 119*, 2005.

[35] A. Kleczkowski and P. Kleczkowski, "Advanced methods for shaping time-frequency areas for the selective mixing of sounds," in *Audio Engineering Society Convention 120*, 2006.

[36] E. Perez Gonzalez and J. D. Reiss, "Automatic mixing: live downmixing stereo panner," in *10th International Conference on Digital Audio Effects (DAFx-07)*, 2007.

[37] E. Perez Gonzalez and J. D. Reiss, "An automatic maximum gain normalization technique with applications to audio mixing," in *Audio Engineering Society Convention 124*, 2008.

[38] E. Perez Gonzalez and J. D. Reiss, "Improved control for selective minimization of masking using inter-channel dependancy effects," in *11th International Conference on Digital Audio Effects (DAFx-08)*, 2008.

[39] E. Perez Gonzalez and J. D. Reiss, "Automatic gain and fader control for live mixing," in *IEEE Workshop on Applications of Signal Processing to Audio and Acoustics (WASPAA)*, 2009.

[40] E. Perez Gonzalez and J. D. Reiss, "Automatic equalization of multichannel audio using cross-adaptive methods," in *Audio Engineering Society Convention 127*, 2009.

[41] M. J. Terrell and J. D. Reiss, "Automatic monitor mixing for live musical performance," *Journal of the Audio Engineering Society*, vol. 57, no. 11, pp. 927–936, 2009.

[42] M. J. Terrell, J. D. Reiss, and M. B. Sandler, "Automatic noise gate settings for drum recordings containing bleed from secondary sources," *EURASIP Journal on Advances in Signal Processing*, vol. 2010, no. 10, 2010.

[43] E. Perez Gonzalez and J. D. Reiss, "A real-time semiautonomous audio panning system for music mixing," *EURASIP Journal on Advances in Signal Processing*, vol. 2010, no. 1, 2010.

[44] J. Scott, M. Prockup, E. M. Schmidt, and Y. E. Kim, "Automatic multi-track mixing using linear dynamical systems," in *8th Sound and Music Computing Conference*, 2011.

[45] J. Scott and Y. E. Kim, "Analysis of acoustic features for automated multi-track mixing," in *12th International Society for Music Information Retrieval Conference (ISMIR 2011)*, 2011.

[46] S. Mansbridge, S. Finn, and J. D. Reiss, "Implementation and evaluation of autonomous multi-track fader control," in *Audio Engineering Society Convention 132*, 2012.

[47] M. J. Terrell and M. B. Sandler, "An offline, automatic mixing method for live music, incorporating multiple sources, loudspeakers, and room effects," *Computer Music Journal*, vol. 36, no. 2, pp. 37–54, 2012.

[48] J. A. Maddams, S. Finn, and J. D. Reiss, "An autonomous method for multi-track dynamic range compression," in *15th International Conference on Digital Audio Effects (DAFx-12)*, 2012.

[49] S. Mansbridge, S. Finn, and J. D. Reiss, "An autonomous system for multitrack stereo pan positioning," in *Audio Engineering Society Convention 133*, 2012.

[50] D. Ward, J. D. Reiss, and C. Athwal, "Multitrack mixing using a model of loudness and partial loudness," in *Audio Engineering Society Convention 133*, 2012.

[51] S. I. Mimilakis, K. Drossos, A. Floros, and D. Katerelos, "Automated tonal balance enhancement for audio mastering applications," in *Audio Engineering Society Convention 134*, 2013.

[52] Z. Ma, J. D. Reiss, and D. A. A. Black, "Implementation of an intelligent equalization tool using Yule-Walker for music mixing and mastering," in *Audio Engineering Society Convention 134*, 2013.

[53] D. Giannoulis, M. Massberg, and J. D. Reiss, "Parameter automation in a dynamic range compressor," *Journal of the Audio Engineering Society*, vol. 61, no. 10, pp. 716–726, 2013.

[54] B. De Man and J. D. Reiss, "A knowledge-engineered autonomous mixing system," in *Audio Engineering Society Convention 135*, 2013.

[55] J. Scott and Y. E. Kim, "Instrument identification informed multi-track mixing," in *14th International Society for Music Information Retrieval Conference (ISMIR 2013)*, 2013.

[56] B. De Man and J. D. Reiss, "Adaptive control of amplitude distortion effects," in *Audio Engineering Society Conference: 53rd International Conference: Semantic Audio*, 2014.

[57] M. J. Terrell, A. Simpson, and M. B. Sandler, "The mathematics of mixing," *Journal of the Audio Engineering Society*, vol. 62, no. 1/2, pp. 4–13, 2014.

[58] P. D. Pestana and J. D. Reiss, "A cross-adaptive dynamic spectral panning technique," in *17th International Conference on Digital Audio Effects (DAFx-14)*, 2014.

[59] M. Hilsamer and S. Herzog, "A statistical approach to automated offline dynamic processing in the audio mastering process," in *17th International Conference on Digital Audio Effects (DAFx-14)*, 2014.

[60] A. Mason, N. Jillings, Z. Ma, J. D. Reiss, and F. Melchior, "Adaptive audio reproduction using personalized compression," in *Audio Engineering Society Conference: 57th International Conference: The Future of Audio Entertainment Technology – Cinema, Television and the Internet*, 2015.

[61] S. Hafezi and J. D. Reiss, "Autonomous multitrack equalization based on masking reduction," *Journal of the Audio Engineering Society*, vol. 63, no. 5, pp. 312–323, 2015.

[62] Z. Ma, B. De Man, P. D. Pestana, D. A. A. Black, and J. D. Reiss, "Intelligent multitrack dynamic range compression," *Journal of the Audio Engineering Society*, vol. 63, no. 6, pp. 412–426, 2015.

[63] D. Matz, E. Cano, and J. Abeßer, "New sonorities for early jazz recordings using sound source separation and automatic mixing tools," in *16th International Society for Music Information Retrieval Conference (ISMIR 2015)*, 2015.

[64] G. Wichern, A. Wishnick, A. Lukin, and H. Robertson, "Comparison of loudness features for automatic level adjustment in mixing," in *Audio Engineering Society Convention 139*, 2015.

[65] E. T. Chourdakis and J. D. Reiss, "Automatic control of a digital reverberation effect using hybrid models," in *Audio Engineering Society Conference: 60th International Conference: DREAMS (Dereverberation and Reverberation of Audio, Music, and Speech)*, 2016.

[66] S. I. Mimilakis, K. Drossos, T. Virtanen, and G. Schuller, "Deep neural networks for dynamic range compression in mastering applications," in *Audio Engineering Society Convention 140*, 2016.

[67] S. I. Mimilakis, E. Cano, J. Abeßer, and G. Schuller, "New sonorities for jazz recordings: Separation and mixing using deep neural networks," in *2nd AES Workshop on Intelligent Music Production*, 2016.

[68] A. Wilson and B. Fazenda, "An evolutionary computation approach to intelligent music production, informed by experimentally gathered domain knowledge," in *2nd AES Workshop on Intelligent Music Production*, 2016.

[69] E. T. Chourdakis and J. D. Reiss, "A machine learning approach to application of intelligent artificial reverberation," *Journal of the Audio Engineering Society*, vol. 65, no. 1/2, pp. 56–65, 2017.

[70] A. L. Benito and J. D. Reiss, "Intelligent multitrack reverberation based on hinge-loss Markov random fields," in *Audio Engineering Society Conference: 2017 AES International Conference on Semantic Audio*, 2017.

[71] F. Everardo, "Towards an automated multitrack mixing tool using answer set programming," in *14th Sound and Music Computing Conference*, 2017.

[72] E. Perez Gonzalez and J. D. Reiss, "Determination and correction of individual channel time offsets for signals involved in an audio mixture," in *Audio Engineering Society Convention 125*, 2008.

[73] B. De Man, J. D. Reiss, and R. Stables, "Ten years of automatic mixing," in *3rd Workshop on Intelligent Music Production*, 2017.

[74] B. De Man, B. Leonard, R. King, and J. D. Reiss, "An analysis and evaluation of audio features for multitrack music mixtures," in *15th International Society for Music Information Retrieval Conference (ISMIR 2014)*, 2014.

[75] E. Deruty, F. Pachet, and P. Roy, "Human–made rock mixes feature tight relations between spectrum and loudness," *Journal of the Audio Engineering Society*, vol. 62, no. 10, pp. 643–653, 2014.

[76] A. Wilson and B. Fazenda, "Variation in multitrack mixes: Analysis of low-level audio signal features," *Journal of the Audio Engineering Society*, vol. 64, no. 7/8, pp. 466–473, 2016.

[77] B. McCarthy, *Sound System Design and Optimisation, Modern Technologies and Tools for Sound Sytem Design and Alignment*. Focal Press, 2007.

[78] J. A. Moorer, "Audio in the new millennium," *Journal of the Audio Engineering Society*, vol. 48, no. 5, pp. 490–498, 2000.

[79] J. Paterson, "What constitutes innovation in music production?," in *Audio Engineering Society Convention 131*, 2011.

[80] J. D. Reiss, "Intelligent systems for mixing multichannel audio," in *17th International Conference on Digital Signal Processing (DSP)*, 2011.

[81] R. Stables, B. De Man, S. Enderby, J. D. Reiss, G. Fazekas, and T. Wilmering, "Semantic description of timbral transformations in music production," in *ACM Multimedia Conference*, 2016.

[82] B. De Man, K. McNally, and J. D. Reiss, "Perceptual evaluation and analysis of reverberation in multitrack music production," *Journal of the Audio Engineering Society*, vol. 65, no. 1/2, pp. 108–116, 2017.

[83] J. Martin, *Programming Real-time Computer Systems*. Prentice-Hall, 1965.

[84] V. Verfaille, C. Guastavino, and C. Traube, "An interdisciplinary approach to audio effect classification," in *9th International Conference on Digital Audio Effects (DAFx-06)*, 2006.

[85] T. Wilmering, G. Fazekas, and M. B. Sandler, "Audio effect classification based on auditory perceptual attributes," in *Audio Engineering Society Convention 135*, 2013.

[86] X. Amatrain, J. Bonada, l. Loscos, J. L. Arcos, and V. Verfaille, "Content-based transformations," *Journal of New Music Research*, vol. 32, no. 1, pp. 95–114, 2003.

[87] U. Zölzer, *DAFX: Digital Audio Effects*. Wiley, 2nd ed., 2011.

[88] J. D. Reiss and A. McPherson, *Audio Effects: Theory, Implementation and Application*. CRC Press, 2014.

[89] V. Verfaille, U. Zolzer, and D. Arfib, "Adaptive digital audio effects (A-DAFx): A new class of sound transformations," *IEEE Transactions on Audio, Speech, and Language Processing*, vol. 14, no. 5, pp. 1817–1831, 2006.

[90] D. Giannoulis, M. Massberg, and J. D. Reiss, "Digital dynamic range compressor design – a tutorial and analysis," *Journal of the Audio Engineering Society*, vol. 60, no. 6, pp. 399–408, 2012.

[91] M. Morrell and J. D. Reiss, "Dynamic panner: An adaptive digital audio effect for spatial audio," in *Audio Engineering Society Convention 127*, 2009.

[92] J. Bitzer and J. LeBoeuf, "Automatic detection of salient frequencies," in *Audio Engineering Society Convention 126*, 2009.

[93] J. Wakefield and C. Dewey, "Evaluation of an algorithm for the automatic detection of salient frequencies in individual tracks of multitrack musical recordings," in *Audio Engineering Society Convention 138*, 2015.

[94] B. Bernfeld, "Attempts for better understanding of the directional stereophonic listening mechanism," in *Audio Engineering Society Convention 44*, 1973.

[95] J. Hodgson, "A field guide to equalisation and dynamics processing on rock and electronica records," *Popular Music*, vol. 29, no. 2, pp. 283–297, 2010.

[96] J. Bitzer, J. LeBoeuf, and U. Simmer, "Evaluating perception of salient frequencies: Do mixing engineers hear the same thing?," in *Audio Engineering Society Convention 124*, 2008.

[97] M. Brandt and J. Bitzer, "Hum removal filters: Overview and analysis," in *Audio Engineering Society Convention 132*, 2012.

[98] M. Mynett, J. P. Wakefield, and R. Till, "Intelligent equalisation principles and techniques for minimising masking when mixing the extreme modern metal genre.," in *Heavy Fundamentalisms: Music, Metal and Politics*, Inter-Disciplinary Press, 2010.

[99] E. Bazil, *Sound Equalization Tips and Tricks*. PC Publishing, 2009.

[100] E. Skovenborg and T. Lund, "Loudness descriptors to characterize programs and music tracks," in *Audio Engineering Society Convention 125*, 2008.

[101] M. Zaunschirm, J. D. Reiss, and A. Klapuri, "A high quality sub-band approach to musical transient modification," *Computer Music Journal*, vol. 36, no. 2, pp. 23–36, 2012.

[102] S. Gorlow and J. D. Reiss, "Model-based inversion of dynamic range compression," *IEEE Transactions on Audio, Speech, and Language Processing*, vol. 21, no. 7, pp. 1434–1444, 2013.

[103] J. Eargle, *The Microphone Book: From Mono to Stereo to Surround – A Guide to Microphone Design and Application*. CRC Press, 2012.

[104] S. Brunner, H.-J. Maempel, and S. Weinzierl, "On the audibility of comb filter distortions," in *Audio Engineering Society Convention 122*, 2007.

[105] P. D. Pestana, J. D. Reiss, and Á. Barbosa, "User preference on artificial reverberation and delay time parameters," *Journal of the Audio Engineering Society*, vol. 65, no. 1/2, pp. 100–107, 2017.

[106] M. R. Schroeder and B. F. Logan, "Colorless artificial reverberation," *IRE Transactions on Audio*, vol. 9, no. 6, pp. 209–214, 1961.

[107] A. Case, *Sound FX: Unlocking the Creative Potential of Recording Studio Effects*. Taylor & Francis, 2012.

[108] R. Izhaki, *Mixing Audio: Concepts, Practices and Tools*. Focal Press, 2008.

[109] V. Välimäki, J. D. Parker, L. Savioja, J. O. Smith, and J. S. Abel, "Fifty years of artificial reverberation," *IEEE Transactions on Audio, Speech, and Language Processing*, vol. 20, no. 5, pp. 1421–1448, 2012.

[110] J. Timoney, V. Lazzarini, B. Carty, and J. Pekonen, "Phase and amplitude distortion methods for digital synthesis of classic analogue waveforms," in *Audio Engineering Society Convention 126*, 2009.

[111] J. Timoney, V. Lazzarini, A. Gibney, and J. Pekonen, "Digital emulation of distortion effects by wave and phase shaping methods," in *13th International Conference on Digital Audio Effects (DAFx-10)*, 2010.

[112] M. Le Brun, "Digital waveshaping synthesis," *Journal of the Audio Engineering Society*, vol. 27, no. 4, pp. 250–266, 1979.

[113] S. Enderby and Z. Baracskai, "Harmonic instability of digital soft clipping algorithms," in *15th International Conference on Digital Audio Effects (DAFx-12)*, 2012.

[114] B. De Man, *Towards a better understanding of mix engineering*. PhD thesis, Queen Mary University of London, 2017.

[115] B. Katz, *Mastering Audio*. Focal Press, 2002.

[116] R. Toulson, "The dreaded mix sign-off," in *Mixing Music*, p. 257, Routledge, 2016.

[117] W. Moylan, *Understanding and Crafting the Mix: The Art of Recording*. Focal Press, 2nd ed., 2006.

[118] A. Case, *Mix Smart: Pro Audio Tips For Your Multitrack Mix*. Focal Press, 2011.

[119] P. D. Pestana, *Automatic Mixing Systems Using Adaptive Digital Audio Effects*. PhD thesis, Universidade Católica Portuguesa, 2013.

[120] N. Jillings and R. Stables, "Investigating music production using a semantically powered digital audio workstation in the browser," in *Audio Engineering Society Conference: 2017 AES International Conference on Semantic Audio*, 2017.

[121] Z. Ma, *Intelligent Tools for Multitrack Frequency and Dynamics Processing*. PhD thesis, Queen Mary University of London, 2016.

[122] P. Harding, "Top-down mixing – A 12-step mixing program," in *Mixing Music*, pp. 62–76, Routledge, 2016.

[123] A. Wilson and B. M. Fazenda, "Navigating the mix-space: Theoretical and practical level-balancing technique in multitrack music mixtures," in *12th Sound and Music Computing Conference*, 2015.

[124] D. Gibson, *The Art Of Mixing: A Visual Guide to Recording, Engineering, and Production*. Thomson Course Technology, 2005.

[125] H. Lee, "Sound source and loudspeaker base angle dependency of phantom image elevation effect," *Journal of the Audio Engineering Society*, vol. 65, no. 9, pp. 733–748, 2017.

[126] H. Haas, "The influence of a single echo on the audibility of speech," *Journal of the Audio Engineering Society*, vol. 20, no. 2, pp. 146–159, 1972.

[127] B. Owsinski, *The Mixing Engineer's Handbook*. Course Technology, 2nd ed., 2006.

[128] S. Stasis, N. Jillings, S. Enderby, and R. Stables, "Audio processing chain recommendation," in *20th International Conference on Digital Audio Effects (DAFx-17)*, 2017.

[129] M. Senior, *Mixing Secrets for the Small Studio: Additional Resources*, Taylor & Francis, 2011.

[130] F. Rumsey, "Mixing and artificial intelligence," *Journal of the Audio Engineering Society*, vol. 61, no. 10, pp. 806–809, 2013.

[131] R. King, B. Leonard, and G. Sikora, "Variance in level preference of balance engineers: A study of mixing preference and variance over time," in *Audio Engineering Society Convention 129*, 2010.

[132] R. King, B. Leonard, and G. Sikora, "Consistency of balance preferences in three musical genres," in *Audio Engineering Society Convention 133*, 2012.

[133] B. Leonard, R. King, and G. Sikora, "The effect of acoustic environment on reverberation level preference," in *Audio Engineering Society Convention 133*, 2012.

[134] B. Leonard, R. King, and G. Sikora, "The effect of playback system on reverberation level preference," in *Audio Engineering Society Convention 134*, 2013.

[135] S. Fenton and H. Lee, "Towards a perceptual model of 'punch' in musical signals," in *Audio Engineering Society Convention 139*, 2015.

[136] A. Wilson and B. M. Fazenda, "Perception of audio quality in productions of popular music," *Journal of the Audio Engineering Society*, vol. 64, no. 1/2, pp. 23–34, 2016.

[137] S. Zagorski-Thomas, "The US vs the UK sound: Meaning in music production in the 1970s," in *The Art of Record Production: An Introductory Reader for a New Academic Field* (S. Frith and S. Zagorski-Thomas, eds.), pp. 57–76, Ashgate, 2012.

[138] H. Massey, *The Great British Recording Studios*. Hal Leonard Corporation, 2015.

[139] B. De Man and J. D. Reiss, "The Mix Evaluation Dataset," in *20th International Conference on Digital Audio Effects (DAFx-17)*, 2017.

[140] A. Pras, B. De Man, and J. D. Reiss, "A case study of cultural influences on mixing practices," in *Audio Engineering Society Convention 144*, 2018.

[141] A. C. Disley and D. M. Howard, "Spectral correlates of timbral semantics relating to the pipe organ," in *Joint Baltic-Nordic Acoustics Meeting*, 2004.

[142] A. Zacharakis, K. Pastiadis, and J. D. Reiss, "An interlanguage study of musical timbre semantic dimensions and their acoustic correlates," *Music Perception: An Interdisciplinary Journal*, vol. 31, no. 4, pp. 339–358, 2014.

[143] R. King, *Recording Orchestra and Other Classical Music Ensembles*. Taylor & Francis, 2016.

[144] J. Scott, "Automated multi-track mixing and analysis of instrument mixtures," in *ACM Multimedia Conference*, 2014.

[145] N. L. Johnson, A. W. Kemp, and S. Kotz, *Univariate Discrete Distributions*, vol. 444. John Wiley & Sons, 2005.

[146] D. Ward, S. Enderby, C. Athwal, and J. D. Reiss, "Real-time excitation based binaural loudness meters," in *18th International Conference on Digital Audio Effects (DAFx-15)*, 2015.

[147] E. Perez Gonzalez and J. D. Reiss, "Automatic mixing," in *DAFX: Digital Audio Effects*, pp. 523–549, Wiley Online Library, 2nd ed., 2011.

[148] A. Clifford and J. D. Reiss, "Calculating time delays of multiple active sources in live sound," in *Audio Engineering Society Convention 129*, 2010.

[149] C. Uhle and J. D. Reiss, "Determined source separation for microphone recordings using IIR filters," in *Audio Engineering Society Convention 129*, 2010.

[150] D. Barchiesi and J. D. Reiss, "Reverse engineering the mix," *Journal of the Audio Engineering Society*, vol. 58, no. 7/8, pp. 563–576, 2010.

[151] E. Perez Gonzalez, *Advanced Automatic Mixing Tools for Music*. PhD thesis, Queen Mary University of London, 2010.

[152] O. Hachour, "The proposed path finding strategy in static unknown environments," *International Journal of Systems Applications, Engineering & Development*, vol. 3, no. 4, pp. 127–138, 2009.

[153] R. Studer, V. R. Benjamins, and D. Fensel, "Knowledge engineering: principles and methods," *Data & Knowledge Engineering*, vol. 25, no. 1–2, pp. 161–197, 1998.

[154] J. Wakefield and C. Dewey, "An investigation into the efficacy of methods commonly employed by mix engineers to reduce frequency masking in the mixing of multitrack musical recordings," in *Audio Engineering Society Convention 138*, 2015.

[155] B. De Man and J. D. Reiss, "A semantic approach to autonomous mixing," *Journal on the Art of Record Production*, vol. 8, 2013.

[156] B. G. Glaser, *Basics of grounded theory analysis: Emergence vs forcing*. Sociology Press, 1992.

[157] A. Pras and C. Guastavino, "The role of music producers and sound engineers in the current recording context, as perceived by young professionals," *Musicae Scientiae*, vol. 15, no. 1, pp. 73–95, 2011.

[158] A. Pras and C. Guastavino, "The impact of producers' comments and musicians' self-evaluation on performance during recording sessions," in *Audio Engineering Society Convention 131*, 2011.

[159] A. Pras and C. Guastavino, "The impact of producers' comments and musicians' self-evaluation on perceived recording quality," *Journal of Music, Technology & Education*, vol. 6, no. 1, pp. 81–101, 2013.

[160] A. Pras, C. Cance, and C. Guastavino, "Record producers' best practices for artistic direction—from light coaching to deeper collaboration with musicians," *Journal of New Music Research*, vol. 42, no. 4, pp. 381–395, 2013.

[161] A. Pras, "What has been left unsaid about studio practices," *Music, Technology, and Education: Critical Perspectives*, pp. 45–62, 2016.

[162] M. A. Martínez Ramírez and J. D. Reiss, "Deep learning and intelligent audio mixing," in *3rd Workshop on Intelligent Music Production*, 2017.

[163] M. A. Martínez Ramírez and J. D. Reiss, "End-to-end equalization with convolutional neural networks," in *21st International Conference on Digital Audio Effects (DAFx-18)*, 2018.

[164] T. Seay, "Primary sources in music production research and education: Using the Drexel University Audio Archives as an institutional model," *Journal of the Art of Record Production*, vol. 5, 2011.

[165] E. Vincent, R. Gribonval, C. Fevotte, A. Nesbit, M. D. Plumbley, M. E. Davies, and L. Daudet, "BASS-dB: The blind audio source separation evaluation database." Available via www.irisa.fr/metiss/BASS-dB, 2010.

[166] J. Abesser, O. Lartillot, C. Dittmar, T. Eerola, and G. Schuller, "Modeling musical attributes to characterize ensemble recordings using rhythmic audio features," in *IEEE International Conference on Acoustics, Speech, and Signal Processing (ICASSP)*, 2011.

[167] A. Liutkus, F.-R. Stöter, Z. Rafii, D. Kitamura, B. Rivet, N. Ito, N. Ono, and J. Fontecave, "The 2016 signal separation evaluation campaign," in *International Conference on Latent Variable Analysis and Signal Separation*, Springer, 2017.

[168] M. Vinyes, "MTG MASS database." www.mtg.upf.edu/static/mass/resources, 2008.

[169] R. M. Bittner, J. Salamon, M. Tierney, M. Mauch, C. Cannam, and J. P. Bello, "MedleyDB: A multitrack dataset for annotation-intensive MIR research," in *15th International Society for Music Information Retrieval Conference (ISMIR 2014)*, 2014.

[170] C.-L. Hsu and J.-S. R. Jang, "On the improvement of singing voice separation for monaural recordings using the MIR-1K dataset," *IEEE Transactions on Audio, Speech, and Language Processing*, vol. 18, no. 2, pp. 310–319, 2010.

[171] Z. Rafii, A. Liutkus, F.-R. Stöter, S. I. Mimilakis, and R. Bittner, "The MUSDB18 corpus for music separation," 2017.

[172] E. Vincent, S. Araki, and P. Bofill, "The 2008 signal separation evaluation campaign: A community-based approach to large-scale evaluation," in *International Conference on Independent Component Analysis and Signal Separation*, Springer, 2009.

[173] "Rock band" (video game). Harmonix, MTV Games, Electronic Arts, 2008.

[174] S. Hargreaves, A. Klapuri, and M. B. Sandler, "Structural segmentation of multitrack audio," *IEEE Transactions on Audio, Speech, and Language Processing*, vol. 20, no. 10, pp. 2637–2647, 2012.

[175] J. Fritsch and M. D. Plumbley, "Score informed audio source separation using constrained nonnegative matrix factorization and score synthesis," in *IEEE International Conference on Acoustics, Speech, and Signal Processing (ICASSP)*, 2013.

[176] R. Bittner, J. Wilkins, H. Yip, and J. P. Bello, "MedleyDB 2.0: New data and a system for sustainable data collection," in *17th International Society for Music Information Retrieval Conference (ISMIR 2016)*, 2016.

[177] B. De Man and J. D. Reiss, "Analysis of peer reviews in music production," *Journal of the Art of Record Production*, vol. 10, 2015.

[178] B. De Man, M. Boerum, B. Leonard, G. Massenburg, R. King, and J. D. Reiss, "Perceptual evaluation of music mixing practices," in *Audio Engineering Society Convention 138*, 2015.

[179] C.-L. Hsu, D. Wang, J.-S. R. Jang, and K. Hu, "A tandem algorithm for singing pitch extraction and voice separation from music accompaniment," *IEEE Transactions on Audio, Speech, and Language Processing*, vol. 20, no. 5, pp. 1482–1491, 2012.

[180] A. Liutkus, R. Badeau, and G. Richard, "Gaussian processes for underdetermined source separation," *IEEE Transactions on Signal Processing*, vol. 59, no. 7, pp. 3155–3167, 2011.

[181] A. Liutkus, F.-R. Stöter, Z. Rafii, D. Kitamura, B. Rivet, N. Ito, N. Ono, and J. Fontecave, "The 2016 signal separation evaluation campaign," in *12th International Conference on Latent Variable Analysis and Signal Separation*, 2017.

[182] Y. Raimond, S. A. Abdallah, M. B. Sandler, and F. Giasson, "The Music Ontology," in *8th International Society for Music Information Retrieval Conference (ISMIR 2007)*, 2007.

[183] K. Arimoto, "Identification of drum overhead-microphone tracks in multi-track recordings," in *2nd AES Workshop on Intelligent Music Production*, 2016.

[184] R. Stables, S. Enderby, B. De Man, G. Fazekas, and J. D. Reiss, "SAFE: A system for the extraction and retrieval of semantic audio descriptors," in *15th International Society for Music Information Retrieval Conference (ISMIR 2014)*, 2014.

[185] T. Zheng, P. Seetharaman, and B. Pardo, "SocialFX: Studying a crowdsourced folksonomy of audio effects terms," in *ACM Multimedia Conference*, 2016.

[186] M. Cartwright and B. Pardo, "Social-EQ: Crowdsourcing an equalization descriptor map," in *14th International Society Music Information Retrieval Conference (ISMIR 2013)*, 2013.

[187] P. Seetharaman and B. Pardo, "Reverbalize: a crowdsourced reverberation controller," in *ACM Multimedia Conference*, 2014.

[188] M. Stein, J. Abeßer, C. Dittmar, and G. Schuller, "Automatic detection of audio effects in guitar and bass recordings," in *Audio Engineering Society Convention 128*, 2010.

[189] T. Bertin-Mahieux, D. P. Ellis, B. Whitman, and P. Lamere, "The Million Song Dataset," in *11th International Conference on Music Information Retrieval (ISMIR 2011)*, 2011.

[190] G. Tzanetakis and P. Cook, "Musical genre classification of audio signals," *IEEE Transactions on Speech and Audio processing*, vol. 10, no. 5, pp. 293–302, 2002.

[191] C. Raffel, *Learning-based methods for comparing sequences, with applications to audio-to-midi alignment and matching*. PhD thesis, Columbia University, 2016.

[192] Z. Duan, B. Pardo, and C. Zhang, "Multiple fundamental frequency estimation by modeling spectral peaks and non-peak regions," *IEEE Transactions on Audio, Speech, and Language Processing*, vol. 18, no. 8, pp. 2121–2133, 2010.

[193] J. Salamon, R. M. Bittner, J. Bonada, J. J. Bosch, E. Gómez, and J. P. Bello, "An analysis/synthesis framework for automatic f0 annotation of multitrack datasets," in *18th International Society for Music Information Retrieval Conference (ISMIR 2017)*, 2017.

[194] C. Donahue, H. H. Mao, and J. McAuley, "The NES music database: A multi-instrumental dataset with expressive performance attributes," in *19th International Society for Music Information Retrieval Conference (ISMIR 2018)*, 2018.

[195] J. Dattorro, "Effect design, part 1: Reverberator and other filters," *Journal of the Audio Engineering Society*, vol. 45, no. 9, pp. 660–684, 1997.

[196] J. Bullock, "LibXtract: A lightweight library for audio feature extraction," in *International Computer Music Conference*, 2007.

[197] Recommendation ITU-R BS.1534-3, "Method for the subjective assessment of intermediate quality level of coding systems," *International Telecommunication Union*, 2003.

[198] C. Völker and R. Huber, "Adaptions for the MUlti Stimulus test with Hidden Reference and Anchor (MUSHRA) for elder and technical unexperienced participants," in *DAGA*, 2015.

[199] M. Schoeffler, F.-R. Stöter, B. Edler, and J. Herre, "Towards the next generation of web-based experiments: A case study assessing basic audio quality following the ITU-R recommendation BS. 1534 (MUSHRA)," in *1st Web Audio Conference*, 2015.

[200] K. J. Woods, M. H. Siegel, J. Traer, and J. H. McDermott, "Headphone screening to facilitate web-based auditory experiments," *Attention, Perception, & Psychophysics*, vol. 79, no. 7, pp. 2064–2072, 2017.

[201] N. Jillings, Y. Wang, J. D. Reiss, and R. Stables, "JSAP: A plugin standard for the Web Audio API with intelligent functionality," in *Audio Engineering Society Convention 141*, 2016.

[202] N. Jillings, J. Bullock, and R. Stables, "JS-Xtract: A realtime audio feature extraction library for the web," in *17th International Society for Music Information Retrieval Conference (ISMIR 2016)*, 2016.

[203] M. R. Genesereth and N. J. Nilsson, *Logical Foundations of Artificial Intelligence*. Morgan Kaufmann Publishers, 1987.

[204] G. Antoniou and F. van Harmelen, "Web Ontology Language: OWL," in *Handbook on Ontologies, International Handbooks on Information Systems* (S. Staab and R. Studer, eds.), pp. 91–110, Springer-Verlag Berlin Heidelberg, 2009.

[205] Y. Raimond, S. A. Abdallah, M. B. Sandler, and F. Giasson, "The Music Ontology," in *8th International Society for Music Information Retrieval Conference (ISMIR 2007)*, 2007.

[206] M.-F. Plassard, ed., *Functional Requirements For Bibliographic Records: final report/IFLA Study Group on the Functional Requirements for Bibliographic Records*, vol. 19. K.G. Saur, 1998.

[207] G. Fazekas and M. B. Sandler, "The Studio Ontology framework," in *12th International Society for Music Information Retrieval Conference (ISMIR 2011)*, 2011.

[208] T. Wilmering, G. Fazekas, and M. B. Sandler, "The Audio Effects Ontology," in *14th International Society for Music Information Retrieval Conference (ISMIR 2013)*, 2013.

[209] T. Wilmering, G. Fazekas, and M. B. Sandler, "AUFX-O: Novel methods for the representation of audio processing workflows," in *International Semantic Web Conference*, 2016.

[210] A. Allik, G. Fazekas, and M. B. Sandler, "Ontological representation of audio features," in *International Semantic Web Conference*, 2016.

[211] S. Enderby, T. Wilmering, R. Stables, and G. Fazekas, "A semantic architecture for knowledge representation in the digital audio workstation," in *2nd AES Workshop on Intelligent Music Production*, 2016.

[212] F. Font, T. Brookes, G. Fazekas, M. Guerber, A. La Burthe, D. Plans, M. D. Plumbley, M. Shaashua, W. Wang, and X. Serra, "Audio Commons: Bringing Creative Commons audio content to the creative industries," in *Audio Engineering Society Conference: 61st International Conference: Audio for Games*, 2016.

[213] M. Ceriani and G. Fazekas, "Audio Commons ontology: A data model for an audio content ecosystem," in *International Semantic Web Conference*, 2018.

[214] S. Bech and N. Zacharov, *Perceptual Audio Evaluation – Theory, Method and Application*. John Wiley & Sons, 2007.

[215] F. E. Toole and S. Olive, "Hearing is believing vs. believing is hearing: Blind vs. sighted listening tests, and other interesting things," in *Audio Engineering Society Convention 97*, 1994.

[216] Recommendation ITU-R BS.1116-3, "Methods for the subjective assessment of small impairments in audio systems," *Radiocommunication Sector of the International Telecommunication Union*, 2015.

[217] S. Olive and T. Welti, "The relationship between perception and measurement of headphone sound quality," in *Audio Engineering Society Convention 133*, 2012.

[218] J. D. Reiss, "A meta-analysis of high resolution audio perceptual evaluation," *Journal of the Audio Engineering Society*, vol. 64, no. 6, pp. 364–379, 2016.

[219] E. Skovenborg, "Development of semantic scales for music mastering," in *Audio Engineering Society Convention 141*, 2016.

[220] B. Cauchi, H. Javed, T. Gerkmann, S. Doclo, S. Goetze, and P. Naylor, "Perceptual and instrumental evaluation of the perceived level of reverberation," in *IEEE International Conference on Acoustics, Speech, and Signal Processing (ICASSP)*, 2016.

[221] B. De Man and J. D. Reiss, "A pairwise and multiple stimuli approach to perceptual evaluation of microphone types," in *Audio Engineering Society Convention 134*, 2013.

[222] J. Francombe, R. Mason, M. Dewhirst, and S. Bech, "Elicitation of attributes for the evaluation of audio-on-audio interference," *Journal of the Acoustical Society of America*, vol. 136, no. 5, pp. 2630–2641, 2014.

[223] A. Pearce, T. Brookes, M. Dewhirst, and R. Mason, "Eliciting the most prominent perceived differences between microphones," *Journal of the Acoustical Society of America*, vol. 139, no. 5, pp. 2970–2981, 2016.

[224] N. Jillings, D. Moffat, B. De Man, and J. D. Reiss, "Web Audio Evaluation Tool: A browser-based listening test environment," in *12th Sound and Music Computing Conference*, 2015.

[225] P. D. Pestana, J. D. Reiss, and A. Barbosa, "Loudness measurement of multitrack audio content using modifications of ITU-R BS.1770," in *Audio Engineering Society Convention 134*, 2013.

[226] S. Olive, "Some new evidence that teenagers and college students may prefer accurate sound reproduction," in *Audio Engineering Society Convention 132*, 2012.

[227] C. Pike, T. Brookes, and R. Mason, "Auditory adaptation to loudspeakers and listening room acoustics," in *Audio Engineering Society Convention 135*, 2013.

[228] B. De Man and J. D. Reiss, "APE: Audio Perceptual Evaluation toolbox for MATLAB," in *Audio Engineering Society Convention 136*, 2014.

[229] J. Berg and F. Rumsey, "Spatial attribute identification and scaling by repertory grid technique and other methods," in *Audio Engineering Society Conference: 16th International Conference: Spatial Sound Reproduction*, 1999.

[230] J. Paulus, C. Uhle, and J. Herre, "Perceived level of late reverberation in speech and music," in *Audio Engineering Society Convention 130*, 2011.

[231] F. Rumsey, "Spatial quality evaluation for reproduced sound: Terminology, meaning, and a scene-based paradigm," *Journal of the Audio Engineering Society*, vol. 50, no. 9, pp. 651–666, 2002.

[232] K. Hermes, T. Brookes, and C. Hummersone, "The influence of dumping bias on timbral clarity ratings," in *Audio Engineering Society Convention 139*, 2015.

[233] J. Berg and F. Rumsey, "Identification of quality attributes of spatial audio by repertory grid technique," *Journal of the Audio Engineering Society*, vol. 54, no. 5, pp. 365–379, 2006.

[234] A. Lindau, V. Erbes, S. Lepa, H.-J. Maempel, F. Brinkman, and S. Weinzierl, "A spatial audio quality inventory (SAQI)," *Acta Acustica united with Acustica*, vol. 100, no. 5, pp. 984–994, 2014.

[235] U. Jekosch, "Basic concepts and terms of 'quality', reconsidered in the context of product-sound quality," *Acta Acustica united with Acustica*, vol. 90, no. 6, pp. 999–1006, 2004.

[236] S. Zielinski, "On some biases encountered in modern listening tests," in *Spatial Audio & Sensory Evaluation Techniques*, 2006.

[237] S. Zielinski, "On some biases encountered in modern audio quality listening tests (part 2): Selected graphical examples and discussion," *Journal of the Audio Engineering Society*, vol. 64, no. 1/2, pp. 55–74, 2016.

[238] F. Rumsey, "New horizons in listening test design," *Journal of the Audio Engineering Society*, vol. 52, no. 1/2, pp. 65–73, 2004.

[239] A. Wilson and B. M. Fazenda, "Relationship between hedonic preference and audio quality in tests of music production quality," in *8th International Conference on Quality of Multimedia Experience (QoMEX)*, 2016.

[240] S. P. Lipshitz and J. Vanderkooy, "The Great Debate: Subjective evaluation," *Journal of the Audio Engineering Society*, vol. 29, no. 7/8, pp. 482–491, 1981.

[241] J. Berg and F. Rumsey, "In search of the spatial dimensions of reproduced sound: Verbal protocol analysis and cluster analysis of scaled verbal descriptors," in *Audio Engineering Society Convention 108*, 2000.

[242] N. Zacharov and T. H. Pedersen, "Spatial sound attributes – development of a common lexicon," in *Audio Engineering Society Convention 139*, 2015.

[243] E. Vickers, "Fixing the phantom center: Diffusing acoustical crosstalk," in *Audio Engineering Society Convention 127*, 2009.

[244] R. King, B. Leonard, and G. Sikora, "The effects of monitoring systems on balance preference: A comparative study of mixing on headphones versus loudspeakers," in *Audio Engineering Society Convention 131*, 2011.

[245] N. Sakamoto, T. Gotoh, and Y. Kimura, "On out-of-head localization in headphone listening," *Journal of the Audio Engineering Society*, vol. 24, no. 9, pp. 710–716, 1976.

[246] F. E. Toole, "The acoustics and psychoacoustics of headphones," in *2nd International Conference of the Audio Engineering Society*, 1984.

[247] M. Schoeffler, F.-R. Stöter, H. Bayerlein, B. Edler, and J. Herre, "An experiment about estimating the number of instruments in polyphonic music: A comparison between internet and laboratory results," in *14th International Society for Music Information Retrieval Conference (ISMIR 2013)*, 2013.

[248] Recommendation ITU-T P.830, "Subjective performance assessment of telephone-band and wideband digital codecs," *International Telecommunication Union*, 1996.

[249] A.-R. Tereping, "Listener preference for concert sound levels: Do louder performances sound better?," *Journal of the Audio Engineering Society*, vol. 64, no. 3, pp. 138–146, 2016.

[250] Recommendation ITU-R BS.1770-4, "Algorithms to measure audio programme loudness and true-peak audio level," *Radiocommunication Sector of the International Telecommunication Union*, 2015.

[251] F. E. Toole, "Listening tests – Turning opinion into fact," *Journal of the Audio Engineering Society*, vol. 30, no. 6, pp. 431–445, 1982.

[252] J. Paulus, C. Uhle, J. Herre, and M. Höpfel, "A study on the preferred level of late reverberation in speech and music," in *Audio Engineering Society Conference: 60th International Conference: DREAMS (Dereverberation and Reverberation of Audio, Music, and Speech)*, 2016.

[253] C. Bussey, M. J. Terrell, R. Rahman, and M. B. Sandler, "Metadata features that affect artificial reverberator intensity," in *Audio Engineering Society Conference: 53rd International Conference: Semantic Audio*, 2014.

[254] W. G. Gardner and D. Griesinger, "Reverberation level matching experiments," in *Sabine Centennial Symposium*, 1994.

[255] J. Berg and F. J. Rumsey, "Validity of selected spatial attributes in the evaluation of 5-channel microphone techniques," in *Audio Engineering Society Convention 112*, 2002.

[256] J. Mycroft, J. D. Reiss, and T. Stockman, "The influence of graphical user interface design on critical listening skills," in *10th Sound and Music Computing Conference*, 2013.

[257] U.-D. Reips, "Standards for internet-based experimenting," *Experimental Psychology*, vol. 49, no. 4, pp. 243–256, 2002.

[258] M. Cartwright, B. Pardo, G. J. Mysore, and M. Hoffman, "Fast and easy crowdsourced perceptual audio evaluation," in *IEEE International Conference on Acoustics, Speech, and Signal Processing (ICASSP)*, 2016.

[259] M. J. Terrell, A. J. R. Simpson, and M. B. Sandler, "A perceptual audio mixing device," in *Audio Engineering Society Convention 134*, 2013.

[260] B. Pardo, D. Little, and D. Gergle, "Building a personalized audio equalizer interface with transfer learning and active learning," in *2nd International ACM Workshop on Music Information Retrieval with User-Centered and Multimodal Strategies*, 2012.

[261] B. Pardo, D. Little, and D. Gergle, "Towards speeding audio EQ interface building with transfer learning," in *International Conference on New Interfaces for Musical Expression (NIME)*, 2012.

[262] A. Sabin and B. Pardo, "Rapid learning of subjective preference in equalization," in *Audio Engineering Society Convention 125*, 2008.

[263] A. Sabin and B. Pardo, "2DEQ: An intuitive audio equalizer," in *ACM Conference on Creativity and Cognition*, 2009.

[264] A. Sabin and B. Pardo, "A method for rapid personalization of audio equalization parameters," in *ACM Multimedia Conference*, 2009.

[265] Z. Rafii and B. Pardo, "Learning to control a reverberator using subjective perceptual descriptors," in *10th International Society for Music Information Retrieval Conference (ISMIR 2009)*, 2009.

[266] P. Seetharaman and B. Pardo, "Crowdsourcing a reverberation descriptor map," in *ACM Multimedia Conference*, 2014.

[267] D. Ward, H. Wierstorf, R. D. Mason, M. Plumbley, and C. Hummersone, "Estimating the loudness balance of musical mixtures using audio source separation," in *3rd Workshop on Intelligent Music Production*, 2017.

[268] N. Jillings and R. Stables, "An intelligent audio workstation in the browser," in *3rd Web Audio Conference*, 2017.

[269] B. C. Moore, B. R. Glasberg, and T. Baer, "A model for the prediction of thresholds, loudness, and partial loudness," *Journal of the Audio Engineering Society*, vol. 45, no. 4, pp. 224–240, 1997.

[270] B. R. Glasberg and B. C. Moore, "A model of loudness applicable to time-varying sounds," *Journal of the Audio Engineering Society*, vol. 50, no. 5, pp. 331–342, 2002.

[271] R. King, B. Leonard, S. Levine, and G. Sikora, "Balance preference testing utilizing a system of variable acoustic condition," in *Audio Engineering Society Convention 134*, 2013.

[272] R. King, B. Leonard, and G. Sikora, "Variance in level preference of balance engineers: A study of mixing preference and variance over time," in *Audio Engineering Society Convention 129*, 2010.

[273] J. Allan and J. Berg, "Evaluation of loudness meters using parameterization of fader movements," in *Audio Engineering Society Convention 135*, 2013.

[274] A. Elowsson and A. Friberg, "Long-term average spectrum in popular music and its relation to the level of the percussion," in *Audio Engineering Society Convention 142*, 2017.

[275] P. D. Pestana, Z. Ma, J. D. Reiss, A. Barbosa, and D. A. Black, "Spectral characteristics of popular commercial recordings 1950–2010," in *Audio Engineering Society Convention 135*, 2013.

[276] E. A. Durant, G. H. Wakefield, D. J. Van Tasell, and M. E. Rickert, "Efficient perceptual tuning of hearing aids with genetic algorithms," *IEEE Transactions on Speech and Audio Processing*, vol. 12, no. 2, pp. 144–155, 2004.

[277] A. C. Neuman, H. Levitt, R. Mills, and T. Schwander, "An evaluation of three adaptive hearing aid selection strategies," *Journal of the Acoustical Society of America*, vol. 82, no. 6, pp. 1967–1976, 1987.

[278] G. H. Wakefield, C. van den Honert, W. Parkinson, and S. Lineaweaver, "Genetic algorithms for adaptive psychophysical procedures: Recipient-directed design of speech-processor maps," *Ear and Hearing*, vol. 26, no. 4, pp. 57S–72S, 2005.

[279] P. G. Stelmachowicz, D. E. Lewis, and E. Carney, "Preferred hearing-aid frequency responses in simulated listening environments," *Journal of Speech, Language, and Hearing Research*, vol. 37, no. 3, pp. 712–719, 1994.

[280] F. K. Kuk and N. M. Pape, "The reliability of a modified simplex procedure in hearing aid frequency response selection," *Journal of Speech, Language, and Hearing Research*, vol. 35, no. 2, pp. 418–429, 1992.

[281] W. Campbell, J. Paterson, and I. van der Linde, "Listener preferences for alternative dynamic-range-compressed audio configurations," *Journal of the Audio Engineering Society*, vol. 65, no. 7/8, pp. 540–551, 2017.

[282] M. Wendl and H. Lee, "The effect of dynamic range compression on loudness and quality perception in relation to crest factor," in *Audio Engineering Society Convention 136*, 2014.

[283] J. Boley, C. Danner, and M. Lester, "Measuring dynamics: Comparing and contrasting algorithms for the computation of dynamic range," in *Audio Engineering Society Convention 129*, 2010.

[284] P. H. Kraght, "Aliasing in digital clippers and compressors," *Journal of the Audio Engineering Society*, vol. 48, no. 11, pp. 1060–1065, 2000.

[285] D. M. Huber and R. Runstein, *Modern Recording Techniques*. Taylor & Francis, 2013.

[286] N. Thiele, "Some thoughts on the dynamics of reproduced sound," *Journal of the Audio Engineering Society*, vol. 53, no. 1/2, pp. 130–132, 2005.

[287] D. M. Ronan, H. Gunes, and J. D. Reiss, "Analysis of the subgrouping practices of professional mix engineers," in *Audio Engineering Society Convention 142*, 2017.

[288] E. Benjamin, "Characteristics of musical signals," in *Audio Engineering Society Convention 97*, 1994.

[289] L. G. Møller, "How much has amplitude distribution of music changed?," in *Audio Engineering Society Convention 71*, 1982.

[290] M. Mijic, D. Masovic, D. Sumarac Pavlovic, and M. Petrovic, "Statistical properties of music signals," in *Audio Engineering Society Convention 126*, 2009.

[291] J. A. McManus, C. Evans, and P. W. Mitchell, "The dynamics of recorded music," in *Audio Engineering Society Convention 95*, 1993.

[292] E. Deruty and D. Tardieu, "About dynamic processing in mainstream music," *Journal of the Audio Engineering Society*, vol. 62, no. 1/2, pp. 42–55, 2014.

[293] M. Oehler, C. Reuter, and I. Czedik-Eysenberg, "Dynamics and low-frequency ratio in popular music recordings since 1965," in *Audio Engineering Society Conference: 57th International Conference: The Future of Audio Entertainment Technology – Cinema, Television and the Internet*, 2015.

[294] M. Kirchberger and F. A. Russo, "Dynamic range across music genres and the perception of dynamic compression in hearing-impaired listeners," *Trends in Hearing*, vol. 20, 2016.

[295] J. Bitzer, D. Schmidt, and U. Simmer, "Parameter estimation of dynamic range compressors: models, procedures and test signals," in *Audio Engineering Society Convention 120*, 2006.

[296] D. Sheng and G. Fazekas, "Automatic control of the dynamic range compressor using a regression model and a reference sound," in *20th International Conference on Digital Audio Effects (DAFx-17)*, 2017.

[297] S. Heise, M. Hlatky, and J. Loviscach, "Automatic adjustment of off-the-shelf reverberation effects," in *Audio Engineering Society Convention 126*, 2009.

[298] A. Czyzewski, "A method of artificial reverberation quality testing," *Journal of the Audio Engineering Society*, vol. 38, no. 3, pp. 129–141, 1990.

[299] Y. Ando, M. Okura, and K. Yuasa, "On the preferred reverberation time in auditoriums," *Acta Acustica united with Acustica*, vol. 50, no. 2, pp. 134–141, 1982.

[300] I. Frissen, B. F. G. Katz, and C. Guastavino, "Effect of sound source stimuli on the perception of reverberation in large volumes," in *Auditory Display: 6th International Symposium*, 2010.

[301] C. Uhle, J. Paulus, and J. Herre, "Predicting the perceived level of late reverberation using computational models of loudness," in *17th International Conference on Digital Signal Processing (DSP)*, 2011.

[302] Z. Meng, F. Zhao, and M. He, "The just noticeable difference of noise length and reverberation perception," in *International Symposium on Communications and Information Technologies (ISCIT'06)*, 2006.

[303] P. Luizard, B. F. Katz, and C. Guastavino, "Perceived suitability of reverberation in large coupled volume concert halls." *Psychomusicology: Music, Mind, and Brain*, vol. 25, no. 3, pp. 317–325, 2015.

[304] A. H. Marshall, D. Gottlob, and H. Alrutz, "Acoustical conditions preferred for ensemble," *Journal of the Acoustical Society of America*, vol. 64, no. 5, pp. 1437–1442, 1978.

[305] S. Hase, A. Takatsu, S. Sato, H. Sakai, and Y. Ando, "Reverberance of an existing hall in relation to both subsequent reverberation time and SPL," *Journal of Sound and Vibration*, vol. 232, no. 1, pp. 149–155, 2000.

[306] M. Barron, "The subjective effects of first reflections in concert halls – The need for lateral reflections," *Journal of Sound and Vibration*, vol. 15, no. 4, pp. 475–494, 1971.

[307] G. A. Soulodre and J. S. Bradley, "Subjective evaluation of new room acoustic measures," *Journal of the Acoustical Society of America*, vol. 98, no. 1, pp. 294–301, 1995.

[308] M. R. Schroeder, D. Gottlob, and K. F. Siebrasse, "Comparative study of European concert halls: Correlation of subjective preference with geometric and acoustic parameters," *Journal of the Acoustical Society of America*, vol. 56, no. 4, pp. 1195–1201, 1974.

[309] D. Lee and D. Cabrera, "Equal reverberance matching of music," in *Proceedings of Acoustics*, 2009.

[310] D. Lee, D. Cabrera, and W. L. Martens, "Equal reverberance matching of running musical stimuli having various reverberation times and SPLs," in *20th International Congress on Acoustics*, 2010.

[311] D. Griesinger, "How loud is my reverberation?," in *Audio Engineering Society Convention 98*, 1995.

[312] W. Kuhl, "Über Versuche zur Ermittlung der günstigsten Nachhallzeit großer Musikstudios," *Acta Acustica united with Acustica*, vol. 4, no. Supplement 2, pp. 618–634, 1954.

[313] E. Kahle and J.-P. Jullien, "Some new considerations on the subjective impression of reverberance and its correlation with objective criteria," in *Sabine Centennial Symposium*, 1994.

[314] Y. Ando, *Concert Hall Acoustics*. Springer-Verlag, 1985.

[315] B. De Man, M. Mora-McGinity, G. Fazekas, and J. D. Reiss, "The Open Multitrack Testbed," in *Audio Engineering Society Convention 137*, 2014.

[316] J. Vanderkooy and K. B. Krauel, "Another view of distortion perception," in *Audio Engineering Society Convention 133*, 2012.

[317] L. Vetter, M. J. Terrell, A. J. R. Simpson, and A. McPherson, "Estimation of overdrive in music signals," in *Audio Engineering Society Convention 134*, 2013.

[318] J. Schimmel, "Using nonlinear amplifier simulation in dynamic range controllers," in *6th International Conference on Digital Audio Effects (DAFx-03)*, 2003.

[319] C. Bönsel, J. Abeßer, S. Grollmisch, and S. Ioannis Mimilakis, "Automatic best take detection for electric guitar and vocal studio recordings," in *2nd AES Workshop on Intelligent Music Production*, 2016.

[320] E. K. Kokkinis, J. D. Reiss, and J. Mourjopoulos, "Detection of 'solo intervals' in multiple microphone multiple source audio applications," in *Audio Engineering Society Convention 130*, 2011.

[321] K. A. Pati and A. Lerch, "A dataset and method for guitar solo detection in rock music," in *Audio Engineering Society Conference: 2017 AES International Conference on Semantic Audio*, 2017.

[322] E. K. Kokkinis, J. D. Reiss, and J. Mourjopoulos, "A Wiener filter approach to microphone leakage reduction in close-microphone applications," *IEEE Transactions on Audio, Speech, and Language Processing*, vol. 20, no. 3, pp. 767–779, 2012.

[323] A. Clifford and J. D. Reiss, "Microphone interference reduction in live sound," in *14th International Conference on Digital Audio Effects (DAFx-11)*, 2011.

[324] A. Clifford and J. D. Reiss, "Reducing comb filtering on different musical instruments using time delay estimation," *Journal of the Art of Record Production*, vol. 5, 2011.

[325] A. Clifford and J. D. Reiss, "Using delay estimation to reduce comb filtering of arbitrary musical sources," *Journal of the Audio Engineering Society*, vol. 61, no. 11, pp. 917–927, 2013.

[326] N. Jillings, A. Clifford, and J. D. Reiss, "Performance optimization of GCC-PHAT for delay and polarity correction under real world conditions," in *Audio Engineering Society Convention 134*, 2013.

[327] P. D. Pestana, J. D. Reiss, and A. Barbosa, "Cross-adaptive polarity switching strategies for optimization of audio mixes," in *Audio Engineering Society Convention 138*, 2015.

[328] M. J. Terrell and J. D. Reiss, "Automatic noise gate settings for multitrack drum recordings," in *12th International Conference on Digital Audio Effects (DAFx-09)*, 2009.

[329] B. H. Deatherage and T. R. Evans, "Binaural masking: Backward, forward, and simultaneous effects," *Journal of the Acoustical Society of America*, vol. 46, no. 2B, pp. 362–371, 1969.

[330] T. R. Agus, M. A. Akeroyd, S. Gatehouse, and D. Warden, "Informational masking in young and elderly listeners for speech masked by simultaneous speech and noise," *Journal of the Acoustical Society of America*, vol. 126, no. 4, pp. 1926–1940, 2009.

[331] M. Geravanchizadeh, H. J. Avanaki, and P. Dadvar, "Binaural speech intelligibility prediction in the presence of multiple babble interferers based on mutual information," *Journal of the Audio Engineering Society*, vol. 65, no. 4, pp. 285–292, 2017.

[332] B. R. Glasberg and B. C. Moore, "Development and evaluation of a model for predicting the audibility of time-varying sounds in the presence of background sounds," *Journal of the Audio Engineering Society*, vol. 53, no. 10, pp. 906–918, 2005.

[333] P. Aichinger, A. Sontacchi, and B. Schneider-Stickler, "Describing the transparency of mixdowns: The masked-to-unmasked-ratio," in *Audio Engineering Society Convention 130*, 2011.

[334] S. Vega and J. Janer, "Quantifying masking in multi-track recordings," in *7th Sound and Music Computing Conference*, 2010.

[335] Z. Ma, J. D. Reiss, and D. A. Black, "Partial loudness in multitrack mixing," in *Audio Engineering Society Conference: 53rd International Conference: Semantic Audio*, 2014.

[336] G. Wichern, H. Robertson, and A. Wishnick, "Quantitative analysis of masking in multitrack mixes using loudness loss," in *Audio Engineering Society Convention 141*, 2016.

[337] N. Jillings and R. Stables, "Automatic masking reduction in balance mixes using evolutionary computing," in *Audio Engineering Society Convention 143*, 2017.

[338] B. C. Moore, B. R. Glasberg, and M. A. Stone, "Why are commercials so loud? Perception and modeling of the loudness of amplitude-compressed speech," *Journal of the Audio Engineering Society*, vol. 51, no. 12, pp. 1123–1132, 2003.

[339] A. Clifford and J. D. Reiss, "Proximity effect detection for directional microphones," in *Audio Engineering Society Convention 131*, 2011.

[340] G. W. Elko, J. Meyer, S. Backer, and J. Peissig, "Electronic pop protection for microphones," in *IEEE Workshop on Applications of Signal Processing to Audio and Acoustics (WASPAA)*, 2007.

[341] L. O'Sullivan and F. Boland, "Towards a fuzzy logic approach to drum pattern humanisation," in *13th International Conference on Digital Audio Effects (DAFx-10)*, 2010.

[342] M. Wright and E. Berdahl, "Towards machine learning of expressive microtiming in Brazilian drumming," in *International Computer Music Conference*, 2006.

[343] F. Gouyon, "Microtiming in samba de roda - preliminary experiments with polyphonic audio," in *Simposio da Sociedade Brasileira de Computacao Musical*, 2007.

[344] L. A. Naveda, F. Gouyon, C. Guedes, and M. Leman, "Multidimensional microtiming in samba music," in *12th Brazilian Symposium on Computer Music*, 2009.

[345] R. Stables, J. Bullock, and I. Williams, "Perceptually relevant models for articulation in synthesised drum patterns," in *Audio Engineering Society Convention 131*, 2011.

[346] R. Stables, S. Endo, and A. Wing, "Multi-player microtiming humanisation using a multivariate Markov model," in *17th International Conference on Digital Audio Effects (DAFx-14)*, 2014.

[347] A. M. Wing, S. Endo, A. Bradbury, and D. Vorberg, "Optimal feedback correction in string quartet synchronization," *Journal of The Royal Society Interface*, vol. 11, no. 93, 2014.

[348] D. Ronan, B. De Man, H. Gunes, and J. D. Reiss, "The impact of subgrouping practices on the perception of multitrack mixes," in *Audio Engineering Society Convention 139*, 2015.

[349] D. Ronan, H. Gunes, D. Moffat, and J. D. Reiss, "Automatic subgrouping of multitrack audio," in *18th International Conference on Digital Audio Effects (DAFx-15)*, 2015.

[350] L. Breiman, "Random forests," *Machine Learning*, vol. 45, no. 1, pp. 5–32, 2001.

[351] N. Jillings and R. Stables, "Automatic channel routing using musical instrument linked data," in *3rd Workshop on Intelligent Music Production*, 2017.

[352] S. Stasis, N. Jillings, S. Enderby, and R. Stables, "Audio processing chain recommendation using semantic cues," in *3rd Workshop on Intelligent Music Production*, 2017.

[353] T. Van Waterschoot and M. Moonen, "Fifty years of acoustic feedback control: State of the art and future challenges," *Proceedings of the IEEE*, vol. 99, no. 2, pp. 288–327, 2011.

[354] D. Barchiesi and J. D. Reiss, "Automatic target mixing using least-squares optimization of gains and equalization settings," in *12th International Conference on Digital Audio Effects (DAFx-09)*, 2009.

[355] M. Ramona and G. Richard, "A simple and efficient fader estimator for broadcast radio unmixing," in *14th International Conference on Digital Audio Effects (DAFx-11)*, 2011.

[356] E. Pampalk, S. Dixon, and G. Widmer, "On the evaluation of perceptual similarity measures for music," in *6th International Conference on Digital Audio Effects (DAFx-03)*, 2003.

[357] S. Gorlow, J. D. Reiss, and E. Duru, "Restoring the dynamics of clipped audio material by inversion of dynamic range compression," in *IEEE International Symposium on Broadband Multimedia Systems and Broadcasting (BMSB)*, 2014.

[358] M. M. Goodwin and C. Avendano, "Frequency-domain algorithms for audio signal enhancement based on transient modification," *Journal of the Audio Engineering Society*, vol. 54, no. 9, pp. 827–840, 2006.

[359] M. Walsh, E. Stein, and J.-M. Jot, "Adaptive dynamics enhancement," in *Audio Engineering Society Convention 130*, 2011.

[360] J. Mycroft, J. D. Reiss, and T. Stockman, "The effect of differing user interface presentation styles on audio mixing," in *International Conference on the Multimodal Experience of Music*, 2015.

[361] T. Page, "Skeuomorphism or flat design: Future directions in mobile device user interface (UI) design education," *International Journal of Mobile Learning and Organisation*, vol. 8, no. 2, pp. 130–142, 2014.

[362] S. Gross, J. Bardzell, and S. Bardzell, "Skeu the evolution: skeuomorphs, style, and the material of tangible interactions," in *8th International Conference on Tangible, Embedded and Embodied Interaction*, 2014.

[363] E. Sanchez, "Skeuominimalism - the best of both worlds." http://edwardsanchez.me/blog/13568587, 2012.

[364] S. Gelineck and S. Serafin, "A quantitative evaluation of the differences between knobs and sliders," in *International Conference on New Interfaces for Musical Expression (NIME)*, 2009.

[365] J. Ratcliffe, "Hand motion-controlled audio mixing interface," *International Conference on New Interfaces for Musical Expression (NIME)*, 2014.

[366] A. Hunt and R. Kirk, "Mapping strategies for musical performance," *Trends in Gestural Control of Music*, pp. 231–258, 2000.

[367] A. Hunt, M. M. Wanderley, and M. Paradis, "The importance of parameter mapping in electronic instrument design," *Journal of New Music Research*, vol. 32, no. 4, pp. 429–440, 2003.

[368] D. Arfib, J.-M. Couturier, L. Kessous, and V. Verfaille, "Strategies of mapping between gesture data and synthesis model parameters using perceptual spaces," *Organised Sound*, vol. 7, no. 2, pp. 127–144, 2002.

[369] E. Schubert and J. Wolfe, "Does timbral brightness scale with frequency and spectral centroid?," *Acta Acustica united with Acustica*, vol. 92, no. 5, pp. 820–825, 2006.

[370] T. Brookes and D. Williams, "Perceptually-motivated audio morphing: Brightness," in *Audio Engineering Society Convention 122*, 2007.

[371] M. Edwards, "Algorithmic composition: Computational thinking in music," *Communications of the ACM*, vol. 54, no. 7, pp. 58–67, 2011.

[372] G. Nierhaus, *Algorithmic composition: Paradigms of Automated Music Generation*. Springer Science & Business Media, 2009.

[373] D. Su, "Using the Fibonacci sequence to time-stretch recorded music," in *3rd Workshop on Intelligent Music Production*, 2017.

[374] S. Stasis, R. Stables, and J. Hockman, "Semantically controlled adaptive equalization in reduced dimensionality parameter space," *Applied Sciences*, vol. 6, p. 116, 2016.

[375] S. Stasis, R. Stables, and J. Hockman, "A model for adaptive reduced-dimensionality equalisation," in *18th International Conference on Digital Audio Effects (DAFx-15)*, 2015.

[376] J. M. Grey, "Multidimensional perceptual scaling of musical timbres," *Journal of the Acoustical Society of America*, vol. 61, no. 5, pp. 1270–1277, 1977.

[377] A. C. Disley, D. M. Howard, and A. D. Hunt, "Timbral description of musical instruments," in *International Conference on Music Perception and Cognition*, 2006.

[378] H. H. Harman, *Modern Factor Analysis*. University of Chicago Press, 1976.

[379] A. Zacharakis, K. Pastiadis, J. D. Reiss, and G. Papadelis, "Analysis of musical timbre semantics through metric and non-metric data reduction techniques," in *12th International Conference on Music Perception and Cognition*, 2012.

[380] B. Schölkopf, A. Smola, and K.-R. Müller, "Nonlinear component analysis as a kernel eigenvalue problem," *Neural Computation*, vol. 10, no. 5, pp. 1299–1319, 1998.

[381] S. Roweis, "EM algorithms for PCA and SPCA," *Advances in Neural Information Processing Systems*, vol. 10, pp. 626–632, 1998.

[382] R. A. Fisher, "The use of multiple measurements in taxonomic problems," *Annals of Eugenics*, vol. 7, no. 2, pp. 179–188, 1936.

[383] L. Van der Maaten and G. Hinton, "Visualizing data using t-SNE," *Journal of Machine Learning Research*, vol. 9, no. 85, pp. 2579–2605, 2008.

[384] G. E. Hinton and R. R. Salakhutdinov, "Reducing the dimensionality of data with neural networks," *Science*, vol. 313, no. 5786, pp. 504–507, 2006.

[385] S. Gelineck, D. M. Korsgaard, and M. Büchert, "Stage vs. channel-strip metaphor: Comparing performance when adjusting volume and panning of a single channel in a stereo mix," in *International Conference on New Interfaces for Musical Expression (NIME)*, 2015.

[386] J. Mycroft, T. Stockman, and J. D. Reiss, "Visual information search in digital audio workstations," in *Audio Engineering Society Convention 140*, 2016.

[387] A. Holladay and B. Holladay, "Audio dementia: A next generation audio mixing software application," in *Audio Engineering Society Convention 118*, 2005.

[388] S. Gelineck and D. Korsgaard, "Stage metaphor mixing on a multi-touch tablet device," in *Audio Engineering Society Convention 137*, 2014.

[389] B. J. Rodriguez Nino and M. Herrera Martinez, "Design of an algorithm for VST audio mixing based on Gibson diagrams," in *Audio Engineering Society Convention 142*, 2017.

[390] W. Gale and J. P. Wakefield, "Investigating the use of virtual reality to solve the underlying problems with the 3D stage paradigm," in *4th Workshop on Intelligent Music Production*, 2018.

[391] S. Gelineck and A. K. Uhrenholt, "Exploring visualisation of channel activity, levels and EQ for user interfaces implementing the stage metaphor for music mixing," in *2nd AES Workshop on Intelligent Music Production*, 2016.

[392] J. Mycroft, J. D. Reiss, and T. Stockman, "A prototype mixer to improve cross-modal attention during audio mixing," in *Audio Mostly Conference*, 2018.

[393] C. Dewey and J. Wakefield, "Formal usability evaluation of audio track widget graphical representation for two-dimensional stage audio mixing interface," in *Audio Engineering Society Convention 142*, 2017.

[394] B. De Man, N. Jillings, and R. Stables, "Comparing stage metaphor interfaces as a controller for stereo position and level," in *4th Workshop on Intelligent Music Production*, 2018.

[395] V. Diamante, "Awol: Control surfaces and visualization for surround creation," Technical report, University of Southern California, Interactive Media Division, 2007.

[396] H. Lee, D. Johnson, and M. Mironovs, "An interactive and intelligent tool for microphone array design," in *Audio Engineering Society Convention 143*, 2017.

[397] H. Lee and F. Rumsey, "Level and time panning of phantom images for musical sources," *Journal of the Audio Engineering Society*, vol. 61, no. 12, pp. 978–988, 2013.

[398] S. Gelineck, J. Andersen, and M. Büchert, "Music mixing surface," in *ACM International Conference on Interactive Tabletops and Surfaces*, 2013.

[399] J. P. Carrascal and S. Jordà, "Multitouch interface for audio mixing," in *International Conference on New Interfaces for Musical Expression (NIME)*, 2011.

[400] S. Jordà, G. Geiger, M. Alonso, and M. Kaltenbrunner, "The reacTable: Exploring the synergy between live music performance and tabletop tangible interfaces," in *1st International Conference on Tangible and Embedded Interaction*, 2007.

[401] S. Gelineck, D. Overholt, M. Büchert, and J. Andersen, "Towards an interface for music mixing based on smart tangibles and multitouch," *International Conference on New Interfaces for Musical Expression (NIME)*, 2013.

[402] N. Liebman, M. Nagara, J. Spiewla, and E. Zolkosky, "Cuebert: A new mixing board concept for musical theatre," in *International Conference on New Interfaces for Musical Expression (NIME)*, 2010.

[403] C. Dewey and J. Wakefield, "Novel designs for the audio mixing interface based on data visualization first principles," in *Audio Engineering Society Convention 140*, 2016.

[404] K. Gohlke, M. Hlatky, S. Heise, D. Black, and J. Loviscach, "Track displays in DAW software: Beyond waveform views," in *Audio Engineering Society Convention 128*, 2010.

[405] M. Cartwright, B. Pardo, and J. D. Reiss, "MIXPLORATION: Rethinking the audio mixer interface," in *International Conference on Intelligent User Interfaces*, 2014.

[406] C. Dewey and J. Wakefield, "Novel designs for the parametric peaking EQ user interface," in *Audio Engineering Society Convention 134*, 2013.

[407] J. Loviscach, "Graphical control of a parametric equalizer," in *Audio Engineering Society Convention 124*, 2008.

[408] S. Heise, M. Hlatky, and J. Loviscach, "A computer-aided audio effect setup procedure for untrained users," in *Audio Engineering Society Convention 128*, 2010.

[409] Z. Chen and G. Hu, "A revised method of calculating auditory exciation patterns and loudness for time-varying sounds," in *IEEE International Conference on Acoustics, Speech, and Signal Processing (ICASSP)*, 2012.

[410] J. Ford, M. Cartwright, and B. Pardo, "MixViz: A tool to visualize masking in audio mixes," in *Audio Engineering Society Convention 139*, 2015.

[411] S. Fenton and J. Wakefield, "Objective profiling of perceived punch and clarity in produced music," in *Audio Engineering Society Convention 132*, 2012.

[412] A. Parker and S. Fenton, "Real-time system for the measurement of perceived punch," in *4th Workshop on Intelligent Music Production*, 2018.

[413] R. I. Godøy and M. Leman, *Musical Gestures: Sound, Movement, and Meaning*. Routledge, 2010.

[414] F. Iazzetta, "Meaning in musical gesture," in *Trends in Gestural Control of Music*, pp. 259–268, 2000.

[415] M. Nielsen, T. B. Moeslund, M. Störring, and E. Granum, "Gesture interfaces," in *HCI Beyond the GUI: Design for Haptic, Speech, Olfactory, and Other Nontraditional Interfaces?* (P. Kortum, ed.), pp. 75–106, Elsevier, 2008.

[416] T. Wilson, S. Fenton, and M. Stephenson, "A semantically motivated gestural interface for the control of a dynamic range compressor," in *Audio Engineering Society Convention 138*, 2015.

[417] W. Westerman, J. G. Elias, and A. Hedge, "Multi-touch: A new tactile 2-d gesture interface for human-computer interaction," in *Human Factors and Ergonomics Society Annual Meeting*, 2001.

[418] W. Balin and J. Loviscach, "Gestures to operate DAW software," in *Audio Engineering Society Convention 130*, 2011.

[419] M. T. Marshall, J. Malloch, and M. M. Wanderley, "Gesture control of sound spatialization for live musical performance," in *Gesture Workshop*, 2007.

[420] M. Lech and B. Kostek, "Testing a novel gesture-based mixing interface," *Journal of the Audio Engineering Society*, vol. 61, no. 5, pp. 301–313, 2013.

[421] J. Ratcliffe, "MotionMix: A gestural audio mixing controller," in *Audio Engineering Society Convention 137*, 2014.

[422] J. Wakefield, C. Dewey, and W. Gale, "LAMI: A gesturally controlled three-dimensional stage leap (motion-based) audio mixing interface," in *Audio Engineering Society Convention 142*, 2017.

[423] R. Selfridge and J. D. Reiss, "Interactive mixing using Wii controller," in *Audio Engineering Society Convention 130*, 2011.

[424] M. J. Morrell, J. D. Reiss, and T. Stockman, "Auditory cues for gestural control of multi-track audio," in *International Community for Auditory Display*, Budapest, Hungary, 2011.

[425] A. L. Fuhrmann, J. Kretz, and P. Burwik, "Multi sensor tracking for live sound transformation," in *International Conference on New Interfaces for Musical Expression (NIME)*, 2013.

[426] P. Modler, "Neural networks for mapping hand gestures to sound synthesis parameters," *Trends in Gestural Control of Music*, pp. 301–314, 2000.

[427] M. Naef and D. Collicott, "A VR interface for collaborative 3D audio performance," in *International Conference on New Interfaces for Musical Expression (NIME)*, 2006.

[428] B. Di Donato, J. Dooley, J. Hockman, J. Bullock, and S. Hall, "MyoSpat: A hand-gesture controlled system for sound and light projections manipulation," in *International Computer Music Conference*, 2017.

[429] K. Drossos, A. Floros, and K. Koukoudis, "Gestural user interface for audio multitrack real-time stereo mixing," in *8th Audio Mostly Conference*, 2013.

[430] T. R. Wilson, *The Gestural Control of Audio Processing*. PhD thesis, University of Huddersfield, 2015.

[431] M. S. O'Modhrain and B. Gillespie, "The moose: A haptic user interface for blind persons," in *Third WWW6 Conference*, 1997.

[432] M. Geronazzo, A. Bedin, L. Brayda, and F. Avanzini, "Multimodal exploration of virtual objects with a spatialized anchor sound," in *Audio Engineering Society Conference: 55th International Conference: Spatial Audio*, 2014.

[433] M. Hlatky, K. Gohlke, D. Black, and J. Loviscach, "Enhanced control of on-screen faders with a computer mouse," in *Audio Engineering Society Convention 126*, 2009.

[434] A. Tanaka and A. Parkinson, "Haptic Wave: A cross-modal interface for visually impaired audio producers," in *CHI Conference on Human Factors in Computing Systems*, 2016.

[435] A. Karp and B. Pardo, "HaptEQ: A collaborative tool for visually impaired audio producers," in *12th Audio Mostly Conference*, 2017.

[436] G. Bradski, "The OpenCV Library," *Dr. Dobb's Journal: Software Tools for the Professional Programmer*, vol. 25, no. 11, pp. 120–123, 2000.

[437] O. Metatla, F. Martin, A. Parkinson, N. Bryan-Kinns, T. Stockman, and A. Tanaka, "Audio-haptic interfaces for digital audio workstations," *Journal on Multimodal User Interfaces*, vol. 10, no. 3, pp. 247–258, 2016.

[438] M. Sarkar, B. Vercoe, and Y. Yang, "Words that describe timbre: A study of auditory perception through language," in *2007 Language and Music as Cognitive Systems Conference*, 2007.

[439] K. Coryat, *Guerrilla Home Recording: How to Get Great Sound from Any Studio (No Matter How Weird or Cheap Your Gear Is)*. MusicPro guides, Hal Leonard Corporation, 2008.

[440] P. White, *Basic Mixers*. The Basic Series, Music Sales, 1999.

[441] M. Cousins and R. Hepworth-Sawyer, *Practical Mastering: A Guide to Mastering in the Modern Studio*. Taylor & Francis, 2013.

[442] G. Waddell, *Complete Audio Mastering: Practical Techniques*. McGraw-Hill Education, 2013.

[443] S. Stasis, J. Hockman, and R. Stables, "Navigating descriptive sub-representations of musical timbre," in *International Conference on New Interfaces for Musical Expression (NIME)*, 2017.

[444] S. Enderby and R. Stables, "A nonlinear method for manipulating warmth and brightness," in *20th International Conference on Digital Audio Effects (DAFx-17)*, 2017.

[445] A. Zacharakis, K. Pastiadis, and J. D. Reiss, "An interlanguage unification of musical timbre," *Music Perception: An Interdisciplinary Journal*, vol. 32, no. 4, pp. 394–412, 2015.

[446] A. Zacharakis, K. Pastiadis, G. Papadelis, and J. D. Reiss, "An investigation of musical timbre: Uncovering salient semantic descriptors and perceptual dimensions.," in *12th International Society for Music Information Retrieval Conference (ISMIR 2011)*, 2011.

[447] S. Wake and T. Asahi, "Sound retrieval with intuitive verbal expressions," in *International Community for Auditory Display*, 1998.

[448] S. Koelsch, "Towards a neural basis of processing musical semantics," *Physics of Life Reviews*, vol. 8, no. 2, pp. 89–105, 2011.

[449] V. Karbusicky, "Grundriss der musikalischen semantik," *Darmstadt: Wissenschaftliche Buchgesellschaft*, 1986.

[450] R. Toulson, "A need for universal definitions of audio terminologies and improved knowledge transfer to the audio consumer," in *2nd Art of Record Production Conference*, 2006.

[451] N. Ford, T. Nind, and F. Rumsey, "Communicating listeners auditory spatial experiences: A method for developing a descriptive language," in *Audio Engineering Society Convention 118*, 2005.

[452] F. Rumsey, *Spatial Audio*. CRC Press, 2012.

[453] W. L. Martens and A. Marui, "Constructing individual and group timbre spaces for sharpness-matched distorted guitar timbres," in *Audio Engineering Society Convention 119*, 2005.

[454] T. H. Pedersen and N. Zacharov, "The development of a sound wheel for reproduced sound," in *Audio Engineering Society Convention 138*, 2015.

[455] M. E. Altinsoy and U. Jekosch, "The semantic space of vehicle sounds: Developing a semantic differential with regard to customer perception," *Journal of the Audio Engineering Society*, vol. 60, no. 1/2, pp. 13–20, 2012.

[456] T. A. Letsche and M. W. Berry, "Large-scale information retrieval with latent semantic indexing," *Information Sciences*, vol. 100, no. 1, pp. 105–137, 1997.

[457] T. Brookes and D. Williams, "Perceptually-motivated audio morphing: Warmth," in *Audio Engineering Society Convention 128*, 2010.

[458] D. Williams and T. Brookes, "Perceptually-motivated audio morphing: Softness," in *Audio Engineering Society Convention 126*, 2009.

[459] A. Zacharakis and J. D. Reiss, "An additive synthesis technique for independent modification of the auditory perceptions of brightness and warmth," in *Audio Engineering Society Convention 130*, 2011.

[460] A. T. Sabin, Z. Rafii, and B. Pardo, "Weighted-function-based rapid mapping of descriptors to audio processing parameters," *Journal of the Audio Engineering Society*, vol. 59, no. 6, pp. 419–430, 2011.

[461] P. Seetharaman and B. Pardo, "Audealize: Crowdsourced audio production tools," *Journal of the Audio Engineering Society*, vol. 64, no. 9, pp. 683–695, 2016.

[462] L. A. Gatys, A. S. Ecker, and M. Bethge, "Image style transfer using convolutional neural networks," in *Proceedings of the IEEE Conference on Computer Vision and Pattern Recognition*, 2016.

[463] S. Dai, Z. Zhang, and G. G. Xia, "Music style transfer: A position paper," arXiv preprint arXiv:1803.06841, 2018.

[464] T. Agus and C. Corrigan, "Adapting audio mixes for hearing-impairments," in *3rd Workshop on Intelligent Music Production*, 2017.

[465] W. Buyens, B. van Dijk, M. Moonen, and J. Wouters, "Music mixing preferences of cochlear implant recipients: A pilot study," *International Journal of Audiology*, vol. 53, no. 5, pp. 294–301, 2014.

[466] G. D. Kohlberg, D. M. Mancuso, D. A. Chari, and A. K. Lalwani, "Music engineering as a novel strategy for enhancing music enjoyment in the cochlear implant recipient," *Behavioural Neurology*, 2015.

[467] V. Pulkki, "Virtual sound source positioning using vector base amplitude panning," *Journal of the Audio Engineering Society*, vol. 45, no. 6, pp. 456–466, 1997.

[468] B. Schmidt, "Interactive mixing of game audio," in *Audio Engineering Society Convention 115*, 2003.

[469] M. Grimshaw, "Sound and immersion in the first-person shooter," *International Journal of Intelligent Games and Simulation*, vol. 5, no. 1, pp. 119–124, 2008.

[470] Mixerman, *Zen and the Art of Mixing*. Hal Leonard Corporation, 2010.

Index